Springer Finance

Editorial Board
M. Avellaneda
G. Barone-Adesi
M. Broadie
M.H.A. Davis
E. Derman
C. Klüppelberg
E. Kopp
W. Schachermayer

Springer Finance

Springer Finance is a programme of books aimed at students, academics and practitioners working on increasingly technical approaches to the analysis of financial markets. It aims to cover a variety of topics, not only mathematical finance but foreign exchanges, term structure, risk management, portfolio theory, equity derivatives, and financial economics.

Ammann M., Credit Risk Valuation: Methods, Models, and Application (2001)
Back K., A Course in Derivative Securities: Introduction to Theory and Computation (2005)
Barucci E., Financial Markets Theory. Equilibrium, Efficiency and Information (2003)
Bielecki T.R. and Rutkowski M., Credit Risk: Modeling, Valuation and Hedging (2002)
Bingham N.H. and Kiesel R., Risk-Neutral Valuation: Pricing and Hedging of Financial Derivatives (1998, 2nd ed. 2004)
Brigo D. and Mercurio F., Interest Rate Models: Theory and Practice (2001)
Buff R., Uncertain Volatility Models-Theory and Application (2002)
Dana R.A. and Jeanblanc M., Financial Markets in Continuous Time (2002)
Deboeck G. and Kohonen T. (Editors), Visual Explorations in Finance with Self-Organizing Maps (1998)
Elliott R.J. and Kopp P.E., Mathematics of Financial Markets (1999, 2nd ed. 2005)
Fengler M., Semiparametric Modeling of Implied Volatility (2005)
Geman H., Madan D., Pliska S.R. and Vorst T. (Editors), Mathematical Finance-Bachelier Congress 2000 (2001)
Gundlach M., Lehrbass F. (Editors), CreditRisk$^+$ in the Banking Industry (2004)
Kellerhals B.P., Asset Pricing (2004)
Külpmann M., Irrational Exuberance Reconsidered (2004)
Kwok Y.-K., Mathematical Models of Financial Derivatives (1998)
Malliavin P. and Thalmaier A., Stochastic Calculus of Variations in Mathematical Finance (2005)
Meucci A., Risk and Asset Allocation (2005)
Pelsser A., Efficient Methods for Valuing Interest Rate Derivatives (2000)
Prigent J.-L., Weak Convergence of Financial Markets (2003)
Schmid B., Credit Risk Pricing Models (2004)
Shreve S.E., Stochastic Calculus for Finance I (2004)
Shreve S.E., Stochastic Calculus for Finance II (2004)
Yor, M., Exponential Functionals of Brownian Motion and Related Processes (2001)
Zagst R., Interest-Rate Management (2002)
Ziegler A., Incomplete Information and Heterogeneous Beliefs in Continuous-time Finance (2003)
Ziegler A., A Game Theory Analysis of Options (2004)
Zhu Y.-L., Wu X., Chern I.-L., Derivative Securities and Difference Methods (2004)

Matthias R. Fengler

Semiparametric Modeling of Implied Volatility

Springer

Matthias R. Fengler
Equity Derivatives Group
Sal. Oppenheim jr. & Cie.
Untermainanlage 1
60329 Frankfurt
Germany
E-mail: matthias.fengler@oppenheim.de

Mathematics Subject Classification (2000): 62G08, 62G05, 62H25
JEL classification: G12, G13

This book is based on the author's dissertation accepted on 28 June 2004 at the Humboldt-Universität zu Berlin.

Library of Congress Control Number: 2005930475

ISBN-10 3-540-26234-2 Springer Berlin Heidelberg New York
ISBN-13 978-3-540-26234-3 Springer Berlin Heidelberg New York

This work is subject to copyright. All rights are reserved, whether the whole or part of the material is concerned, specifically the rights of translation, reprinting, reuse of illustrations, recitation, broadcasting, reproduction on microfilm or in any other way, and storage in data banks. Duplication of this publication or parts thereof is permitted only under the provisions of the German Copyright Law of September 9, 1965, in its current version, and permission for use must always be obtained from Springer. Violations are liable for prosecution under the German Copyright Law.

Springer is a part of Springer Science+Business Media
springeronline.com
© Springer-Verlag Berlin Heidelberg 2005
Printed in The Netherlands

The use of general descriptive names, registered names, trademarks, etc. in this publication does not imply, even in the absence of a specific statement, that such names are exempt from the relevant protective laws and regulations and therefore free for general use.

Typesetting: by the author and TechBooks using a Springer LaTeX macro package

Cover design: *design & production,* Heidelberg

Printed on acid-free paper   SPIN: 11496786   41/TechBooks   5 4 3 2 1 0

Le Monde Instable

Le monde en vne isle porté
Sur la mer tant esmeue et rogue,
Sans seur gouuernal nage et vogue,
Monstrant son instabilité.

Corrozet (1543)

quoted from Henkel and Schöne (1996)

Acknowledgements

This book has benefitted a lot from suggestions and comments of colleagues, fellow students and friends whom I wish to thank at this place. At first rate, I thank Wolfgang Härdle. He directed my interest to implied volatilities and made me familiar with non- and semiparametric modeling in Finance. Without him, his encouragement and advise this work would not exist. Furthermore, I like to thank Vladimir Spokoiny, in particular for his comments during my talks in the Seminar for Mathematical Statistics at the WIAS, Berlin.

This work is in close context with essays I have written with a number of coauthors. Above all, I thank Enno Mammen: the cooperation in semiparametric modeling has been highly instructive and fruitful for me. In this regard, I also thank Qihua Wang.

For an unknown number of helpful discussions or proofreading my thanks go to Peter Bank, Michal Benko, Szymon Borak, Kai Detlefsen, Erhard and Martin Fengler, Patrick Herbst, Zdeněk Hlávka, Torsten Kleinow, Danilo Mercurio and Marlene Müller and to all contemporary and former members of the ISE and CASE for the inspiring working environment they generated there.

Finally, I wish to thank the members of my family non explicitly mentioned up to now, Stephanus and especially my mother Brigitte Fengler and Georgia Mavrodi who in their ways did all their best to support me and the project at its different stages.

I gratefully acknowledge financial support by the Deutsche Forschungsgemeinschaft in having been a member of the Sonderforschungsbereich 373 *Quantifikation und Simulation ökonomischer Prozesse* at the Humboldt-Universität zu Berlin.

Berlin, May 2005   Matthias R. Fengler

Frequently Used Notation

Abbreviation or symbol	Explanation
ATM	at-the-money
BS	Black and Scholes (1973)
cdf	cumulative distribution function
C_t	price of a call option at time t
C_t^{BS}	Black-Scholes price of a call option at time t
$\mathcal{C}(\mathcal{A})$	the continuous functions $f : \mathcal{A} \to \mathbb{R}$
$\mathcal{C}^k(\mathcal{A})$	functions in $\mathcal{C}(\mathcal{A})$ with continuous derivatives up to order k
$\mathcal{C}^{k,l}(\mathbb{R} \times \mathbb{R})$	the functions $f : \mathbb{R} \times \mathbb{R} \to \mathbb{R}$ which are $\mathcal{C}^k$ w.r.t. the first and $\mathcal{C}^l$ w.r.t. the second argument
$\mathsf{Cov}(X,Y)$	covariance of two random variables X and Y
CPC(A)	common principal component (analysis)
δ	dividend yield
δ_{x_0}	Dirac delta function defined by the property: $\int f(x)\,\delta_{x_0}(x)\,dx = f(x_0)$ for a smooth function f
$\mathsf{E}(X)$	expected value of the random variable X
F_t	forward or futures price of an asset at time t
$\mathcal{F}_t$	filtration, the information set generated by the information available up to time t
$\mathbf{I}_p$	$p \times p$ unity matrix
IV	implied volatility
IVS	implied volatility surface
ITM	in-the-money
$\mathbf{1}(\mathcal{A})$	indicator function of the set $\mathcal{A}$
K	exercise price

X Notations

$K(\cdot)$	kernel function: continuous, bounded and symmetric real function satisfying $\int K(u)\, du = 1$
κ_f	forward or futures moneyness: $\kappa_f \stackrel{\text{def}}{=} K/F_t$
LVS	local volatility surface
μ	mean of a random variable
$N(\boldsymbol{\mu}, \boldsymbol{\Sigma})$	normal distribution with mean vector $\boldsymbol{\mu}$ and covariance matrix $\boldsymbol{\Sigma}$
OTM	out-of-the-money
o	$\alpha_n = o(\beta_n)$ means: $\lim_{n\to\infty} \frac{\alpha_n}{\beta_n} \to 0$
$\mathcal{O}$	$\alpha_n = \mathcal{O}(\beta_n)$ means: $\lim_{n\to\infty} \frac{\alpha_n}{\beta_n} \to$ some constant
pdf	probability density function
pCPC(q)	partial CPC model of order q
P_t	price of a put option at time t
PCA	principal component analysis
$P(\mathcal{A})$	probability of the set $\mathcal{A}$, objective measure
PDE	partial differential equation
Q	a risk neutral measure
r	interest rate
$\mathbb{R}^d$	d-dimensional Euclidian space, $\mathbb{R} = \mathbb{R}^1$
$\mathbb{R}^+$	the non-negative real numbers
S_t	price of a stock at time t
SDE	stochastic differential equation
$\boldsymbol{\Sigma}$	covariance matrix
t	time
T	expiry date of a financial contract
τ	$\tau \stackrel{\text{def}}{=} T - t$, time to maturity of an option or a forward
Var(X)	variance of the random variable X
W_t	Brownian motion at time t
$\overline{W}_t$	Brownian motion under the risk neutral measure at time t
$\varphi(x)$	pdf of the normel distribution: $\varphi(x) \stackrel{\text{def}}{=} \frac{1}{\sqrt{2\pi}} e^{-x^2/2}$
$\Phi(u)$	cdf of a normal random variable: $\Phi(u) \stackrel{\text{def}}{=} \int_{-\infty}^{u} \frac{1}{\sqrt{2\pi}} e^{-x^2/2}\, dx$
$\stackrel{\text{def}}{=}$	is defined as
$\sim$	if $X \sim D$, the random variable X has the distribution D
$\stackrel{\mathcal{L}}{\to}$	converges in distribution to
$\stackrel{P}{\to}$	converges in probability to
$(X)^+$	$(X)^+ \stackrel{\text{def}}{=} \max(X, 0)$
$\langle X \rangle_t$	quadratic variation process of the stochastic process X
$\langle X, Y \rangle_t$	covariation process of the stochastic processes X and Y
$|x|$	absolute value of the scalar x
$|\mathbf{X}|$	determinant of the matrix $\mathbf{X}$
$\mathbf{X}^\top$	transpose of the matrix $\mathbf{X}$
tr $\mathbf{X}$	trace of the matrix $\mathbf{X}$
$\langle f, g \rangle$	inner product of the functions f and g

In this book, we will mainly employ three concepts of volatility based on the following stochastic differential equation for the asset price process:

$$\frac{dS_t}{S_t} = \mu(S_t, t)\, dt + \sigma(S_t, t, \cdot)\, dW_t .$$

These concepts are in particular:

Instantaneous	Implied	Local
	—— volatility ——	
$\sigma(S_t, t, \cdot)$	$\widehat{\sigma}_t(K, T)$	$\sigma_{K,T}(S_t, t)$
Instantaneous volatility measures the instantaneous standard deviation of the return process of the log-asset price. It depends on the current level of the asset price S_t, time t and possibly on other state variables abbreviated with '$\cdot$'.	*Implied* volatility is the BS option price implied measure of volatility. It is the volatility parameter that equates the BS price and a particular observed market price of an option. Thus, it depends on the strike K, the expiry date T and time t.	*Local* volatility is the expected instantaneous volatility conditional on a particular level of the asset price $S_T = K$ at $t = T$. If the instantaneous volatility is a deterministic function in S_t and t, i.e. can be written as $\sigma(S_t, t)$, then $\sigma_{K,T}(S_t, t) = \sigma(K, T)$.

The term **volatility** is reserved for objects of the kind σ and $\widehat{\sigma}$, while their squared counterparts σ^2 and $\widehat{\sigma}^2$ are called **variance**.

Contents

1 Introduction 1

2 The Implied Volatility Surface 9
   2.1 The Black-Scholes Model 9
   2.2 The Self-Financing Replication Strategy 11
   2.3 Risk Neutral Pricing 12
   2.4 The BS Formula and the Greeks 14
   2.5 The IV Smile 19
   2.6 Static Properties of the Smile Function 27
      2.6.1 Bounds on the Slope 27
      2.6.2 Large and Small Strike Behavior 28
   2.7 General Regularities of the IVS 30
      2.7.1 Static Stylized Facts 30
      2.7.2 DAX Index IV between 1995 and 2001 33
   2.8 Relaxing the Constant Volatility Case 34
      2.8.1 Deterministic Volatility 35
      2.8.2 Stochastic Volatility 36
   2.9 Challenges Arising from the Smile 40
      2.9.1 Hedging and Risk Management 40
      2.9.2 Pricing 42
   2.10 IV as Predictor of Realized Volatility 42
   2.11 Why Do We Smile? 43
   2.12 Summary 46

3 Smile Consistent Volatility Models 47
   3.1 Introduction 47
   3.2 The Theory of Local Volatility 49
   3.3 Backing the LVS Out of Observed Option Prices 51
   3.4 The *dual* PDE Approach to Local Volatility 54
   3.5 From the IVS to the LVS 55
   3.6 Asymptotic Relations Between Implied and Local Volatility ... 60

Contents

- 3.7 The Two-Times-IV-Slope Rule for Local Volatility 62
- 3.8 The K-Strike and T-Maturity Forward Risk-Adjusted Measure 64
- 3.9 Model-Free (Implied) Volatility Forecasts 66
- 3.10 Local Volatility Models 67
  - 3.10.1 Deterministic Implied Trees 67
  - 3.10.2 Stochastic Implied Trees 80
  - 3.10.3 Reconstructing the LVS 84
- 3.11 Excellent Fit, but...: the Delta Problem 88
- 3.12 Stochastic IV Models 91
- 3.13 Summary 94

4 Smoothing Techniques 97
- 4.1 Introduction 97
- 4.2 Nadaraya-Watson Smoothing 99
  - 4.2.1 Kernel Functions 99
  - 4.2.2 The Nadaraya-Watson Estimator 100
- 4.3 Local Polynomial Smoothing 102
- 4.4 Bandwidth Selection 104
  - 4.4.1 Theoretical Framework 104
  - 4.4.2 Bandwidth Choice in Practice 106
- 4.5 Least Squares Kernel Smoothing 115
  - 4.5.1 The LSK Estimator of the IVS 115
  - 4.5.2 Application of the LSK Estimator 117
- 4.6 Summary 123

5 Dimension-Reduced Modeling 125
- 5.1 Introduction 125
- 5.2 Common Principal Component Analysis 128
  - 5.2.1 The Family of CPC Models 128
  - 5.2.2 Estimating Common Eigenstructures 131
  - 5.2.3 Stability Tests for Eigenvalues and Eigenvectors 134
  - 5.2.4 CPC Model Selection 138
  - 5.2.5 Empirical Results 139
- 5.3 Functional Data Analysis 155
  - 5.3.1 Basic Set-Up of FPCA 156
  - 5.3.2 Computing FPCs 157
- 5.4 Semiparametric Factor Models 160
  - 5.4.1 The Model 162
  - 5.4.2 Norming of the Estimates 166
  - 5.4.3 Choice of Model Parameters 167
  - 5.4.4 Empirical Analysis 171
  - 5.4.5 Assessing Prediction Performance 182
- 5.5 Summary 184

6	**Conclusion and Outlook**	187
A	**Description and Preparation of the IV Data**	189
	A.1 Preliminaries	189
	A.2 Data Correction Scheme	190
B	**Some Results from Stochastic Calculus**	195
C	**Proofs of the Results on the LSK IV Estimator**	201
	C.1 Proof of Consistency	201
	C.2 Proof of Asymptotic Normality	203
References		207
Index		221

1

Introduction

> *Yet that weakness is also its greatest strength. People like the model because they can easily understand its assumptions. The model is often good as a first approximation, and if you can see the holes in the assumptions you can use the model in more sophisticated ways.*
>
> Black (1992)

Expected volatility as a measure of risk involved in economic decision making is a key ingredient in modern financial theory: the rational, risk-averse investor will seek to balance the tradeoff between the risk he bears and the return he expects. The more volatile the asset is, i.e. the more it is prone to excessive price fluctuations, the higher will be the expected premium he demands. Markowitz (1959), followed by Sharpe (1964) and Lintner (1965), were among the first to quantify the idea of the simple equation 'more risk means higher return' in terms of equilibrium models. Since then, the analysis of volatility and price fluctuations has sparked a vast literature in theoretical and quantitative finance that refines and extends these early models. As the most recent climax of this story, one may see the Nobel prize in Economics granted to Robert Engle in 2003 for his path-breaking work on modeling time-dependent volatility.

Long before this, a decisive turn in the research of volatility was rendered possible with the seminal publication by Black and Scholes (1973) on the pricing of options and corporate liabilities. Their fundamental result, the celebrated Black-Scholes (BS) formula, offers a framework for the valuation of European style derivatives within a simple set of assumptions. Six parameters enter the pricing formula: the current underlying asset price, the strike price, the expiry date of the option, the riskless interest rate, the dividend yield, and a constant volatility parameter that describes the instantaneous standard deviation of the returns of the log-asset price. The application of the formula, however, faces an obstacle: only its first five parameters are known quantities. The last one, the volatility parameter, is not.

An obvious way to respond to this dilemma is to resort to well-established statistical tools and to estimate the volatility parameter from the time series data of the underlying asset. However, there is also a second perspective that the markets and the literature quickly adopted: instead of estimating the volatility for finding an option price, one aims at recovering that volatility

1 Introduction

IVS Ticks 20000502

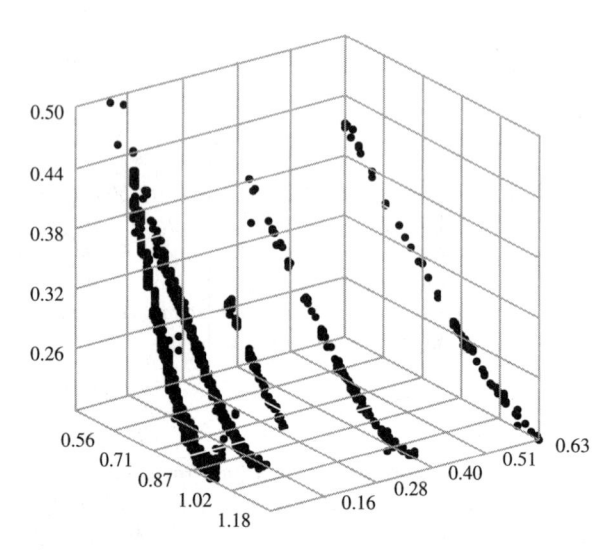

Fig. 1.1. DAX option IVs on 20000502. IV observations are displayed as *black* dots. *Lower left* axis is moneyness and *lower right* time to maturity measured in years

which the market has priced into a given option price observation. To put it in other words, the question is:

what volatility is implied in observed option prices, if the BS model is a valid description of market conditions?

This *reverse* perspective constitutes the concept of the *BS implied volatility*.

A typical picture of implied volatility (IV), as observed on 2nd May, 2000, or 20000502 (a date notation we will adopt from now on) is presented in Fig. 1.1. IV is displayed across different strike prices and expiry dates. Strikes are rescaled in a moneyness metric, where strikes near the current asset price are mapped into the neighborhood of one, and the expiry dates are converted into the time to maturity of the option expressed in years. As is visible, IV exhibits a pronounced curvature across strikes and is also curved across time to maturity, albeit not so much. For a given time to maturity, this function has been named *smile*, and the entire ensemble is called the *implied volatility surface* (IVS). The striking conclusion from a picture like Fig. 1.1 is the clear contradiction to an assumption fundamental to the BS model: instead of being constant, IV is nonlinear in strikes and time to maturity, and – if seen in a sequel of points in time – also time-dependent.

This evident antagonism has been a fruitful starting-point for variations and extensions of this basic pricing model in any direction. At the same time, it does not appear to harm the model itself or the popularity of IV. Nowadays, IV is ubiquitous: it serves as a convenient way of quoting options among market participants, volatility trading is common practice on trading floors, market models incorporate the risk from fluctuating IVs for hedges, and risk management tools, which are approved by banking regulators to steer the allocation of economic capital, include models of the IVS.

A number of reasons may be put forward for explaining the unrivalled popularity of IV. One of them – already anticipated by the initial words by Fisher Black – can be seen in the set of easy-to-communicate assumptions associated with the BS model. Another, more fundamental reason is that a volatility concept implied from option prices enjoys a particular – if not pivotal – property: as options are bets on the future development of the underlying asset, the key advantage of this *option implied volatility* is the fact that it is a forward looking variable by nature. Thus, unlike volatility measures based on historical data, it should reflect *market expectations* on volatility over the remaining life time of the option. Consequently, the information content of IV and its capability of being a predictor for future asset price volatility has been of primary concern in the literature on IV from the early studies up to now.

Yet, it was only in the recent decade that the finance community recognized that the IVS – aside from being a potential predictor or well-known artefact and curiosity – bears valuable information on the asset price process and its dynamics, and that this information can be exploited in models for the pricing and the hedging of other complex derivatives or positions. This development goes in line with the advent of highly liquid option and futures markets that were established all around the world beginning from the nineteen-nineties. Before this, model calibration and pricing typically relied on historically sampled time series data. This bears the disadvantage that the results are predominantly determined by the price history and that the adjustment to new information is too slow. Unlike time series data, the cross-sectional dimension of option prices across different strikes over a range of time to maturities offers the unique opportunity to directly exploit *instantaneous* data for model calibration.

This breakthrough, initiated by the work of Derman and Kani (1994a), Dupire (1994) and Rubinstein (1994), triggered the literature on *smile consistent pricing*. It led, for instance, to the development of static option replication as a means of hedging or to implied trees as a pricing tool. The challenge for this new approach is that IV cannot be directly used as an input factor, since – as shall be seen in the course of this book – IV is a *global* measure of volatility. Pricing requires a *local* measure of volatility. Hence, at the heart of this theory there is another volatility concept, called *local volatility*. Local volatility, unfortunately, cannot be observed and needs to be extracted from market data, either from option prices or from the IVS. Other modeling approaches formulate IV as an additional stochastic process, that – together with the asset price process – enters the pricing equation of derivatives.

These developments explain why the new focus actuated the interest in refined modeling techniques of the IVS and in the structural analysis of its dynamics. In modeling the IVS, one faces two principal challenges: as is visible from Fig. 1.1, the estimators are required to provide sufficient *functional flexibility* in order to optimally fit the shape of the IVS. Otherwise, a model bias will ensue. Second, given the high-dimensional complexity of the IVS, *low-dimensional representations* are desirable from a dynamic standpoint. Not only does a low-dimensional representation of the IVS facilitate the practical implementation of any (dynamic) model, it additionally uncovers the structural basis of the data. This will ultimately lead to a better understanding of the IVS as a financial variable. Natural candidates of techniques that meet these key requirements are *non- and semiparametric methods*: they allow for high functional flexibility and parsimonious modeling. Therefore, results from this line of research are of immediate importance when local volatility or stochastic IV models are to be implemented in practice.

The aim of this book is twofold: the first object is to give a thorough treatment of the financial theory on implied and local volatility and smile consistent modeling. Particular attention is given to highlight the cross-relationships between the volatility concepts as shown in Fig. 1.1. The second object is to familiarize the reader with refined non- and semiparametric estimation strategies and dimension reduction methods for functional surfaces and to demonstrate their effectiveness in the field of IV modeling. The majority of results and techniques we discuss are currently available in preprints or published papers, only. In having their applicability in mind, we take care to illustrate them with empirical investigations that underline their use in practice. We believe that in combining the two fields of research – smile consistent modeling and non- and semiparametric estimation techniques – in this way, we can fill a gap among the textbooks at today's disposal.

Writing a book in the mid of two fields of research requires concessions to the breadth each topic can be treated with. Since our emphasis is on financial modeling aspects, we introduce both financial and statistical theory to the extent we deem necessary for the reader to fully appreciate the core concepts of the book. At the same time, we try to keep the book as self-contained as possible in providing an appendix that collects main results from stochastic calculus and statistics. Therefore, general asset pricing theory is introduced only in its basics. For a broader and more general overview on asset pricing theory the reader is referred to classical textbooks such as Björk (1998), Duffie (2001), Föllmer and Schied (2002), Hull (2002), Joshi (2003), or Lipton (2001) to name but a few. The same philosophy applies to the non- and semiparametric methods. Standard books the reader may like to consult in this direction are provided, e.g., by Efromovich (1999), Härdle (1990), Härdle et al. (2004), Horowitz (1998), Pagan and Ullah (1999), and Ramsay and Silverman (1997).

Local volatility models or their stochastic ramifications are not the only way to price derivatives. Of same significance are approaches relying on stochastic volatility specifications and on Lévy processes. Indeed, the current literature on derivatives pricing may be divided into two main camps: the partisans of local volatility models who prefer them, because local volatility models produce an almost excellent fit to the observed option data; and those who criticize local volatility models principally for predicting the wrong smile dynamics. It is this second camp that favors stochastic volatility specifications and Lévy models. In this book, we enter the particulars of this debate, but topics like stochastic volatility and Lévy models are only briefly touched. In doing so, we do not intend to argue that these competing modeling approaches are not justified: they certainly are, and there are very good arguments in favor of them. Rather it is our intention to bring together this important strand of literature and to discuss advantages and potential drawbacks. The pricing of derivatives in stochastic volatility models can be found in the excellent textbooks by Fouque et al. (2000) and Lewis (2000), and an outstanding treatment of jump diffusions is provided in Cont and Tankov (2004), or in Schoutens (2003).

Many computations for this book were done in XploRe. XploRe is a software which provides a combination of classical and modern statistical procedures together with sophisticated, interactive graphics. XploRe also allows for web-based computing services. Therefore this text is offered as an e-book, i.e. it is designed as an interactive document with links to other features. The e-book may be downloaded from www.xplore-stat.de using the license key given on the last page of this book. The e-book design offers a PDF and HTML file with links to MD*Tech computing servers.

Organization of the Book

In Chap. 2, we give an introduction into the classical BS model. The basic option valuation techniques are presented to derive the celebrated BS pricing formula. Next, the concepts of IV and the IVS are introduced. Given the model's inconsistency with the empirical evidence, potential directions of relaxing the rigid assumptions are discussed. This will lead to new interpretations of IV as *averages of volatility*. We proceed in discussing the consequences that arise for pricing and hedging in the presence of the smile. A short summary of the literature that investigates IV as a predictor of realized volatility follows. The chapter concludes by giving an account of the potential reasons for the existence of a non-constant smile function.

Chapter 3 is devoted to local volatility. Up to now, the theoretical relationship between implied and local volatility – and finally instantaneous volatility as the measure of the contemporaneous asset price variability – is not as clearcut as one might wish. In certain boundary situations or asymptotic regimes only has it been possible to make the relation more precise. Figure 1.2 gives an overview of the current state of research. All relations are developed in the course of the next two chapters. The relationship, possibly most important from a practical point of view, is presented by the dotted line, linking implied and local volatility. It represents the so called *IV counterpart of the Dupire formula*, which enables the pricing of exotic options directly from an estimate of the IVS and its derivatives. The chapter discusses several methods to extract local volatility, especially implied tree techniques. Implied trees can be

6   1 Introduction

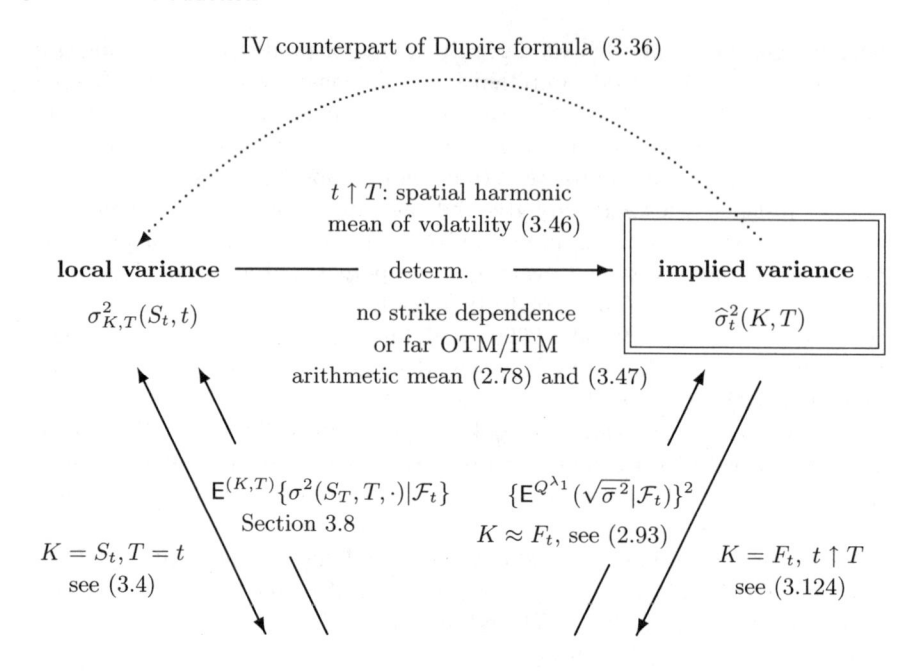

Fig. 1.2. Overview on the volatility concepts important to this work. *Solid lines* denote exact concepts about how the different types of volatility are linked. The *dotted line* represents an ad-hoc relationship. The *arrows* denote the direction of the relation. The term volatility is reserved for objects of the kind σ and $\widehat{\sigma}$, while their squared counterparts σ^2 and $\widehat{\sigma}^2$ are called variance

considered as nonparametric approximations to the local volatility function. The so called delta debate of local volatility models is covered. The chapter concludes by presenting the class of stochastic IV models.

In Chap. 4, we move to smoothing techniques of the IVS. We introduce the Nadaraya-Watson estimator as the simplest nonparametric estimator for the IVS. This is followed by local polynomial estimation, which is decisive when it comes to the estimation of derivatives. Finally, we introduce a least squares kernel estimator of the IVS. The least squares kernel estimator smoothes the IVS in the space of option prices and avoids the potentially undesirable two-step procedure of previous estimators: traditionally, in the first step, implied volatilities are derived. In the second step the actual fitting algorithm is applied. A two-step estimator may be less biased, when option prices or other input parameters can be observed with errors, only.

The probably biggest challenge in IVS modeling is dimension reduction. This is the topic of Chap. 5, which is divided into two major parts. The first part, focusses on linear transformations of the IVS. A standard approach in statistics is to apply principal component analysis. In principal component analysis the high-dimensional variables are projected into a lower dimensional space such that as little information as possible is lost. However, this approach is not directly applicable to the IVS due to the surface structure. Hence, we use the *common principal component models* that we find to allow for a parsimonious, yet flexible model choice. A concern of applying the principal component transformation is stability across time. We derive and apply stability tests across different annual samples. The first part concludes by modeling the resulting factors via standard GARCH time series techniques.

The second part of Chap. 5 is devoted to nonlinear transformations via functional principal component techniques. We first outline the functional principal component framework. Then we propose a semiparametric factor model for the IVS. The semiparametric factor model provides a number of advantages compared with other methods: first, surface estimation *and* dimension reduction can be achieved in one single step. Second, it estimates in the local neighborhood of the design points of the surface, only. With regard to Fig. 1.1 this means that we estimate only in the local vicinity of the black dots. This will avoid model biases. Third, the technique delivers a small set of functions and factor loadings that span the propagation of the IVS through space and time. We provide another time series analysis of these factors based on vector autoregressive models and perform a horse race which compares the model against a simpler practitioners' model.

Chapter 6 concludes and gives directions to future research.

2

The Implied Volatility Surface

> *A smiley implied volatility is the wrong number to put in the wrong formula to obtain the right price.*
>
> Rebonato (1999)

2.1 The Black-Scholes Model

The option pricing model developed by Black and Scholes (1973) and further extended by Merton (1973) is a landmark in financial theory. It laid the foundations of preference-free valuation of contingent claims. Despite its rather restrictive assumptions and the large number of refinements to the model available today, it remains an important benchmark and cornerstone of financial model building. Here, we give a short review of the BS model and present the fundamental results necessary for the further development of this work. For a more detailed account, we refer to textbooks in Finance, such as Musiela and Rutkowski (1997) or Karatzas (1997).

We consider a continuous-time economy with a trading interval $[0, T^*]$, where $T^* > 0$. It is assumed that trading can take place continuously, that there are no differences between lending and borrowing rates, no taxes and short-sale constraints.

Let $(\Omega, \mathcal{F}, P)$ be a probability space, and $(W_t)_{0 \leq t \leq T^*}$ a Brownian motion (see appendix Chap. B for a definition of the Brownian motion) defined on this space. P is the objective probability measure. Information in the economy is revealed by a filtration $(\mathcal{F}_t)_{0 \leq t \leq T^*}$, which is the P-augmentation of the natural filtration

$$\mathcal{F}_t^W = \sigma(W_s,\ 0 \leq s \leq t),\quad 0 \leq t \leq T^*\ . \tag{2.1}$$

The filtration is assumed to satisfy the 'usual' conditions, namely that it is right-continuous, and that $\mathcal{F}_0$ contains all null sets.

The asset price $(S_t)_{0 \leq t \leq T^*}$, which pays a constant dividend yield δ, is modelled by a geometric Brownian motion adapted to $(\mathcal{F}_t)_{0 \leq t \leq T^*}$. The evolution of the asset is given by the stochastic differential equation (SDE):

$$\frac{dS_t}{S_t} = \mu\, dt + \sigma\, dW_t\ , \tag{2.2}$$

where μ denotes the (constant) instantaneous drift and σ the (constant) instantaneous (or spot) volatility function. The quantity σ^2 measures the instantaneous variance of the return process of $\ln S_t$. Thus, instantaneous volatility

σ can be interpreted as the (local) measure of the risk incurred when investing one monetary unit into the risky asset, Frey (1996).

The solution to the SDE (2.2) is given by

$$S_t = S_0 \exp\left\{\left(\mu - \frac{1}{2}\sigma^2\right)t + \sigma W_t\right\}, \quad \forall t \in [0, T^*], \qquad (2.3)$$

where $S_0 > 0$. This is seen from applying the Itô formula, given in (B.10), to (2.3). Since (2.3) is a functional of the Brownian Motion W_t, it is a strong solution; for the precise conditions, conditions guaranteeing uniqueness and existence of a solution to (2.2), see in appendix Chap. B.

The economy is endowed with a savings account or riskless bond with constant interest rate r, which is described by the ordinary differential equation:

$$dB_t = rB_t\,dt, \qquad (2.4)$$

with boundary condition $B_0 \stackrel{\text{def}}{=} 1$, or equivalently $B_t = e^{rt}$, for all $t \in [0, T^*]$.

An option, also called derivative or contingent claim, is a security whose payoff depends on a primary asset, such as the stock price. This asset is usually referred to as the *underlying asset*. For instance, a *call option* entitles the buyer the right – but not the obligation – to buy the underlying asset for a known price K, the *exercise price*. A *put option* entitles the buyer the right to sell the underlying asset for a known price K. We say that an option is of *European style* if it can only be exercised at a prespecified expiry date $T \leq T^*$. If the option can be exercised at any date $t \in [0, T]$ during its life time, the option is said to be of *American style*.

At the maturity date T, the value of a European call contract is given by the payoff function

$$\psi(S_T) = (S_T - K)^+, \qquad (2.5)$$

where $(S_T - K)^+ \stackrel{\text{def}}{=} \max(S_T - K, 0)$. For a put option the payoff is:

$$\psi(S_T) = (K - S_T)^+. \qquad (2.6)$$

These simple derivatives are also called *plain vanilla* options. They are nowadays tradable as standardized contracts on almost any futures exchange market around the world.

In order to receive a payoff such as (2.5) and (2.6), the investor must pay an option price, or *option premium*, to a counterparty when the contract is entered. The investor is also said to be *long* in the option, while the counterparty has a *short* position. The counterparty is obliged to deliver the payoff according to the prespecified conditions. In any case, also when the option expires worthless, the short position earns the option premium paid initially by the long side. Option theory deals with finding this option premium, i.e. it is about the valuation, or the pricing of contingent claims.

There are two important methodologies for deriving the prices of contingent claims: first, a replication strategy based on a self-financing portfolio that provides the same terminal payoff as the derivative. By no-arbitrage considerations, the capital necessary for setting up this portfolio must equal the price of the derivative. Second, there is a probabilistic approach which computes the derivative price as the discounted expectation of the payoff under an equivalent martingale measure (so called risk neutral measure). Both strategies will be sketched in the following.

2.2 The Self-Financing Replication Strategy

A trading strategy is given by a pair of progressively measurable processes $(a_t)_{0 \leq t \leq T}$ and $(b_t)_{0 \leq t \leq T}$, which denote the number of shares held in the stock and the amount of money stored in the savings account. They must satisfy $P\left(\int_0^T a_t^2 dt < \infty\right) = 1$ and $P\left(\int_0^T |b_t| dt < \infty\right) = 1$ such that the stochastic and usual integrals involving a_t and b_t are well defined. Denote the portfolio value by $V_t = a_t S_t + b_t B_t$.

We say that there is an *arbitrage opportunity* in the market if $(V_t)_{0 \leq t \leq T}$ satisfies for $V_0 = 0$:
$$V_T \geq 0 \text{ and } P(V_T > 0) > 0 , \qquad (2.7)$$
In words: if there were an arbitrage opportunity in the market, we would finish in T from zero capital with positive probability of gain at no risk.

The portfolio is called to be *self-financing*, if it satisfies:
$$\begin{aligned} dV_t &= a_t \, dS_t + b_t \, dB_t + a_t \delta S_t \, dt \\ &= a_t(\mu + \delta) S_t \, dt + a_t \sigma S_t \, dW_t + b_t r B_t \, dt , \end{aligned} \qquad (2.8)$$
since the stock pays a dividend $\delta S_t dt$ within the small interval dt. Self-financing means that gains and losses in the portfolio are entirely due to changes in the stock and the bond.

It should be remarked that the self-financing property is not sufficient to exclude arbitrage opportunities. Additionally it is required that the value process $(V_t)_{0 \leq t \leq T}$ has a finite lower bound: it is called to be *tame*, Karatzas (1997).

The price of a contingent claim is a function denoted by $H(S_t, t)$. It shall be assumed that $H \in \mathcal{C}^{2,1}(\mathbb{R}^+ \times (0,T))$, i.e. it is contained in the set of functions which are twice in their first and once in their second argument continuously differentiable. The portfolio replicates the contingent claim if for some pair $(a_t)_{0 \leq t \leq T}$ and $(b_t)_{0 \leq t \leq T}$:
$$V_t = a_t S_t + b_t B_t = H(S_t, t) , \quad \forall t \in [0, T] . \qquad (2.9)$$

Applying the Itô formula (B.10) to $H(S_t, t)$ yields:
$$\begin{aligned} dH(S_t, t) &= \frac{\partial H}{\partial t} dt + \frac{\partial H}{\partial S} dS_t + \frac{1}{2} \frac{\partial^2 H}{\partial S^2} d\langle S \rangle_t \\ &= \left(\frac{\partial H}{\partial t} + \mu S_t \frac{\partial H}{\partial S} + \frac{1}{2} \sigma^2 S_t^2 \frac{\partial^2 H}{\partial S^2} \right) dt + \sigma S_t \frac{\partial H}{\partial S} dW_t . \end{aligned} \qquad (2.10)$$

The quadratic variation process $\langle X \rangle_t$ has a t-subscript in order to distinguish it from our notation for the inner product $\langle \cdot, \cdot \rangle$, which is introduced in Sect. 5.3.

Equating the coefficients of (2.8) and (2.10) in the dW_t terms shows:

$$a_t = \frac{\partial H}{\partial S} . \qquad (2.11)$$

From the replication condition (2.9), the trading strategy in the bond is obtained as

$$b_t = e^{-rt}\left\{H(S_t, t) - S_t \frac{\partial H}{\partial S}\right\} . \qquad (2.12)$$

With these results of a_t and b_t, equate the coefficients of dt-terms in (2.8) and (2.10). This shows:

$$0 = \frac{\partial H}{\partial t} + (r - \delta)S\frac{\partial H}{\partial S} + \frac{1}{2}\sigma^2 S^2 \frac{\partial^2 H}{\partial S^2} - rH . \qquad (2.13)$$

Thus, the price of any European option has to satisfy this partial differential equation (henceforth: BS PDE) with the appropriate boundary condition $H(S_T, T) = \psi(S_T)$. The solution to (2.13) is the value of the replicating portfolio. In fact, for any payoff function $\psi(x)$ continuous on $\mathbb{R}$, for which the condition $\int_{-\infty}^{+\infty} e^{-\alpha x^2}|\psi(x)|dx < \infty$ holds for some $\alpha > 0$, derivative prices can be found by solving (2.13), Musiela and Rutkowski (1997).

The remarkable feature of this result is that pricing the derivative within this model is independent of the appreciation rate μ. Thus market participants may have a different idea about the appreciation rate of the stock, they will agree on the derivative price as long as they agree on the other parameters in the model. This result is closely related to the uniqueness of a risk neutral measure to be introduced next.

2.3 Risk Neutral Pricing

The idea about risk neutral pricing is to introduce a new probability measure Q such that the discounted value process $\widetilde{V}_t \stackrel{\text{def}}{=} e^{-rt}V_t$ of the replicating portfolio is a martingale under Q, i.e. it satisfies:

$$\widetilde{V}_t = \mathsf{E}^{\mathsf{Q}}(\widetilde{V}_T|\mathcal{F}_t) , \qquad (2.14)$$

see Appendix B. By the *fundamental theorem of asset pricing*, originally due to Harrison and Kreps (1979), such a measure exists *if and only if* the market is arbitrage-free.

Since the portfolio replicates the derivative, i.e. $V_T = \psi(S_T)$, (2.14) implies

$$V_t = \mathsf{E}^{\mathsf{Q}}\{e^{-r(T-t)}\psi(S_T)|\mathcal{F}_t\} , \qquad (2.15)$$

which is, by the arguments in Sect. 2.1, the price of the derivative. This means that under the measure Q, pricing derivatives is reduced to computing (conditional) expectations.

The change of measure is achieved by Girsanov's Theorem, which is given in the technical appendix in Chap. B. It states that there exists a measure Q equivalent to P (i.e. both measures agree on the same null sets) such that the discounted stock price is a martingale under Q. In our setting, since we have a continuous dividend payment, we require the discounted process with cumulative dividend re-investments $\hat{S}_t \stackrel{\text{def}}{=} e^{\delta t} S_t$ to be the martingale. It is denoted by $\widetilde{S}_t \stackrel{\text{def}}{=} e^{-rt} \hat{S}_t$.

By Girsanov's Theorem the new measure Q is computed via the Radon-Nikodým derivative P almost surely

$$\frac{dQ}{dP} = \exp\left\{-\lambda W_{T^*} - \frac{1}{2}\lambda^2 T^*\right\}, \qquad (2.16)$$

where

$$\lambda \stackrel{\text{def}}{=} \frac{\mu + \delta - r}{\sigma}. \qquad (2.17)$$

Under Q the discounted price process $\widetilde{S}_t$ satisfies

$$d\widetilde{S}_t = \sigma \widetilde{S}_t \, d\overline{W}_t, \qquad (2.18)$$

where

$$\overline{W}_t \stackrel{\text{def}}{=} W_t + \lambda t, \quad \forall t \in [0, T^*], \qquad (2.19)$$

is a Brownian motion on the space $(\Omega, \mathcal{F}, Q)$. The object λ is called *market price of risk*, since it measures the excess return $\mu + \delta - r$ per unit of risk borne by the investor. The term vanishes under Q, whence the name risk neutral pricing. The risk neutral measure is unique if and only if the market is complete, i.e., every integrable contingent claim can be replicated by a tame portfolio. This is the case, when the number of tradable assets and the number of driving Brownian motions coincide, Karatzas (1997, Theorem 0.3.5).

That $(\widetilde{V}_t)_{0 \leq t \leq T}$ is indeed a martingale is seen by the following manipulations:

$$\begin{aligned}
d\widetilde{V}_t &= -re^{-rt} V_t \, dt + e^{-rt} \, dV_t \\
&= -re^{-rt}(a_t e^{-\delta t}\hat{S}_t + b_t B_t)\, dt + e^{-rt}(a_t e^{-\delta t}\, d\hat{S}_t + b_t \, dB_t) \\
&= a_t e^{-\delta t} d(e^{-rt}\hat{S}_t) \\
&= a_t e^{-\delta t} d\widetilde{S}_t \\
&= a_t e^{-\delta t} \sigma \widetilde{S}_t \, d\overline{W}_t, \qquad (2.20)
\end{aligned}$$

where the second step follows from (2.8) and (2.9) rewritten in terms of $\hat{S}_t$.

The BS pricing formula for a plain vanilla call is found by computing

$$\mathsf{E}^Q\{\psi(S_T - K)^+ | \mathcal{F}_t\}. \qquad (2.21)$$

This is done by noting that

$$\ln(\widetilde{S}_T/\widetilde{S}_t) \sim N\left(r - \delta - \frac{1}{2}\sigma^2(T-t), \sigma^2(T-t)\right), \tag{2.22}$$

where $N(\mu, \sigma^2)$ is the normal distribution with mean μ and variance σ^2. The solution is given in the following section.

2.4 The BS Formula and the Greeks

The price $C(S_t, t)$ of a plain vanilla call is the solution to the PDE (2.13) with the boundary condition $C(S_T, T) = (S_T - K)^+$. The explicit solution is known as the Black and Scholes (1973) formula for calls:

$$C^{BS}(S_t, t, K, T, \sigma, r, \delta) = e^{-\delta\tau} S_t \Phi(d_1) - e^{-r\tau} K \Phi(d_2), \tag{2.23}$$

where

$$d_1 = \frac{\ln(S_t/K) + (r - \delta + \frac{1}{2}\sigma^2)\tau}{\sigma\sqrt{\tau}}, \tag{2.24}$$

$$d_2 = d_1 - \sigma\sqrt{\tau}, \tag{2.25}$$

and where $\Phi(u) \stackrel{\text{def}}{=} \int_{-\infty}^{u} \varphi(x)\,dx$ is the cdf of the standard normal distribution, whose pdf is given by $\varphi(x) \stackrel{\text{def}}{=} \frac{1}{\sqrt{2\pi}} e^{-x^2/2}$ for $x \in \mathbb{R}$. S_t denotes the asset price at time t. K is the strike or exercise price (the notation is not to be confused with the kernel functions denoted by $K(\cdot)$ in Chap. 4). The expiry date of the option is T, and $\tau \stackrel{\text{def}}{=} T - t$ denotes its time to maturity. As in (2.2), σ is the constant volatility function. The riskless interest rate is denoted by r, and the constant dividend yield by δ. It is easy to check using the relevant derivatives given in (2.28) to (2.37) that the BS price satisfies the BS PDE (2.13).

To clarify notation: we will rarely enumerate all parameters of an option pricing function $C(S_t, t, K, T, \sigma, r, \delta)$ explicitly. Rather we limit the enumeration to those parameters that are important for the exposition in the certain context. Sometimes we find it convenient to simply denote the time dependence as a t-subscript: C_t.

The price of a put option $P(S_t, t, K, T, \sigma, r, \delta)$ on the same asset with same expiry and same strike price, which has the payoff function $\psi(S_T) = (K - S_T)^+$, can be obtained from the *put-call parity*:

$$C_t - P_t = e^{-\delta\tau} S_t - e^{-r\tau} K. \tag{2.26}$$

This is a model-free relationship that follows from the trivial fact that $S_T - K = (S_T - K)^+ - (K - S_T)^+$. The BS put price is found to be:

$$P^{BS}(S_t, t, K, T, \sigma, r, \delta) = e^{-r\tau} K \Phi(-d_2) - e^{-\delta\tau} S_t \Phi(-d_1), \tag{2.27}$$

where d_1 and d_2 are defined as in (2.24) and (2.25).

2.4 The BS Formula and the Greeks

Delta

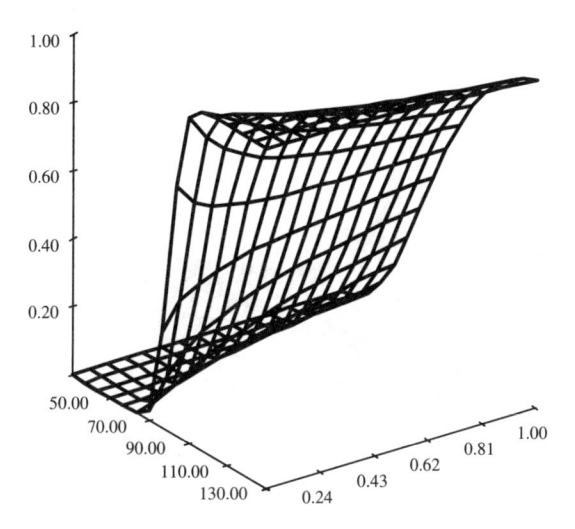

Fig. 2.1. Call delta (2.28) as a function of asset prices (*left axes*) and time to maturity (*right axes*) for $K = 100$ ◨ SCMdelta.xpl

In hedging and risk management, but also for the further exposition of this book, the derivatives of the BS formula, the so called *greeks*, play an important role. In the following, we present their formulae together with the names commonly used on the trading floor, and shortly discuss the most important ones. For some of these derivatives – to the best of our knowledge – there do not exist any nicknames. Usually, these are sensitivities of less immediate concern in daily practice, such as the derivatives with respect to the strike price. A more detailed discussion of the properties and the use of the greeks can be found in Hull (2002) and Franke et al. (2004).

The first derivative with respect to the stock price, the *delta*, gives the number of shares of the underlying asset to be held in the hedge portfolio. This was shown in Equation (2.11). From (2.28), but also from the example in Fig. 2.1, it is seen that the delta for a call is positive throughout. Thus, the replication portfolio is always long in the stock. Of equal importance in practice is the *gamma* (2.29), which measures the convexity of the price function in the stock. The gamma achieves its maximum in the neighborhood of the current asset price, Fig. 2.2. From the put-call parity (2.26) it is seen that the put gamma and the call gamma are equivalent.

The *vega* (2.30), which is the option's sensitivity to changes in volatility, is plotted in Fig. 2.3. It is seen that it increases for longer time to maturities. Put and call vega are equal. The second derivative with respect to volatility (2.31), which is termed *volga*, is displayed in Fig. 2.4. For the strikes in the neighborhood of the current asset price it is typically very low and negative. *Theta*

16   2 The Implied Volatility Surface

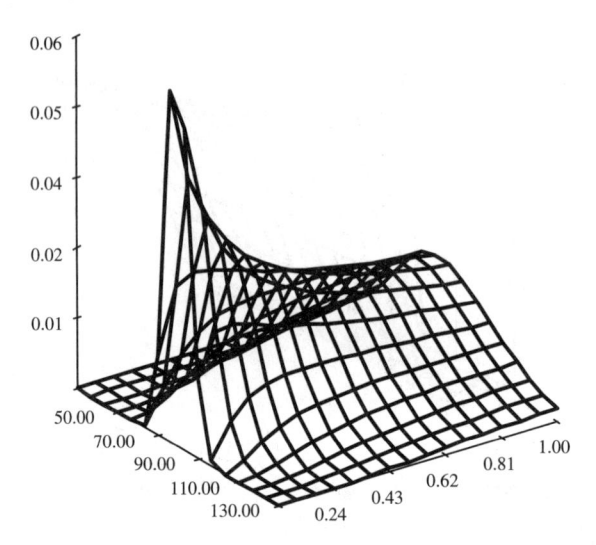

Gamma

Fig. 2.2. Gamma (2.29) as a function of asset prices (*left axes*) and time to maturity (*right axes*) for $K = 100$ SCMgamma.xpl

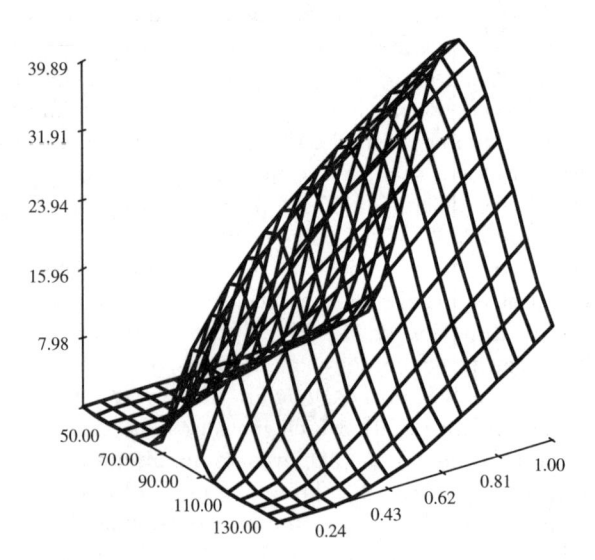

Vega

Fig. 2.3. Vega (2.30) as a function of asset prices (*left axes*) and time to maturity (*right axes*) for $K = 100$ SCMvega.xpl

2.4 The BS Formula and the Greeks

Volga

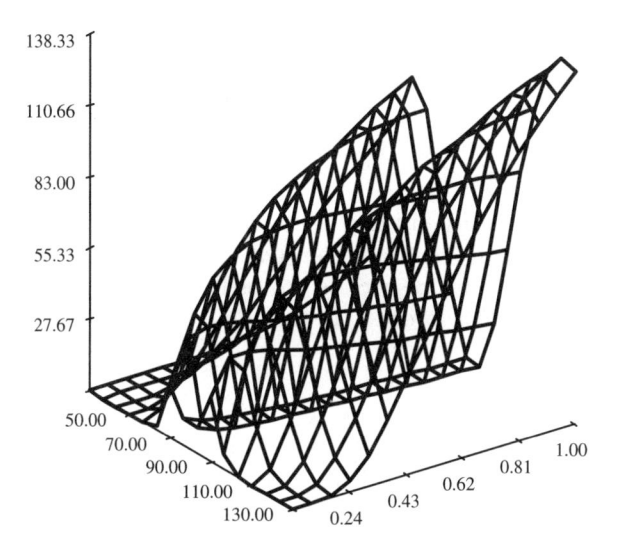

Fig. 2.4. Volga (2.31) as a function of asset prices (*left axes*) and time to maturity (*right axes*) for $K = 100$ 🔘 SCMvolga.xpl

measures the sensitivity of the option to time decay, and *rho* is the sensitivity with respect to interest rate changes.

The formulae of the greeks are given by:

$$\textbf{delta} \quad \frac{\partial C_t}{\partial S} = e^{-\delta\tau}\Phi(d_1) \tag{2.28}$$

$$\textbf{gamma} \quad \frac{\partial^2 C_t}{\partial S^2} = \frac{e^{-\delta\tau}\varphi(d_1)}{S_t \sigma \sqrt{\tau}} \tag{2.29}$$

$$\textbf{vega} \quad \frac{\partial C_t}{\partial \sigma} = e^{-\delta\tau} S_t \sqrt{\tau} \varphi(d_1) \tag{2.30}$$

$$\textbf{volga} \quad \frac{\partial^2 C_t}{\partial \sigma \partial \sigma} = e^{-\delta\tau} S_t \sqrt{\tau} \varphi(d_1) \frac{d_1 d_2}{\sigma} \tag{2.31}$$

$$\textbf{vanna} \quad \frac{\partial^2 C_t}{\partial \sigma \partial S} = -e^{-\delta\tau} \varphi(d_1) \frac{d_2}{\sigma} \tag{2.32}$$

$$\frac{\partial C_t}{\partial K} = -e^{-r\tau}\Phi(d_2) \tag{2.33}$$

$$\frac{\partial^2 C_t}{\partial K^2} = \frac{e^{-r\tau}\varphi(d_2)}{\sigma\sqrt{\tau}K} = \frac{e^{-\delta\tau}S_t\varphi(d_1)}{\sigma\sqrt{\tau}K^2} \tag{2.34}$$

$$\frac{\partial^2 C_t}{\partial \sigma \partial K} = \frac{e^{-\delta\tau} S_t d_1 \varphi(d_1)}{\sigma K} \tag{2.35}$$

theta
$$\frac{\partial C_t}{\partial t} = -\frac{\partial C_t}{\partial T}$$
$$= -\frac{e^{-\delta\tau}S_t\sigma\varphi(d_1)}{2\sqrt{\tau}}$$
$$+ \delta e^{-\delta\tau}S_t\Phi(d_1) - re^{-r\tau}K\Phi(d_2) \quad (2.36)$$

rho
$$\frac{\partial C_t}{\partial r} = \tau e^{-r\tau}K\Phi(d_2) \quad (2.37)$$

An important quantity among the 'unnamed' greeks is the second derivative of the option with respect to the strike price (2.34). The reason is that the second derivative with respect to the strike price yields the *risk neutral (transition) density* of the process. In the empirical literature it is also called *state price density*:

$$\phi(K,T|S_t,t) \stackrel{\text{def}}{=} e^{r(T-t)}\frac{\partial^2 C_t(K,T)}{\partial K^2} . \quad (2.38)$$

The probability that the stock arrives at levels $K \in [K_1, K_2]$ at date T, given that the stock is at level S_t in t, is computed by:

$$Q(S_T \in [K_1, K_2]) = \int_{K_1}^{K_2} \phi(K,T|S_t,t)\,dK . \quad (2.39)$$

Relationship (2.34) yields the specific BS transition density as:

$$\phi(K,T|S_t,t) = \frac{\varphi(d_2)}{\sigma\sqrt{\tau}K} , \quad (2.40)$$

which is a log-normal pdf in K. Of course, this is just another way to see that

$$\ln S_T \sim N\left(\ln S_t + \left(r - \delta - \frac{1}{2}\sigma^2\right)\tau, \sigma^2\tau\right) , \quad (2.41)$$

as was explained earlier.

The second derivative with respect to the strike, however, is useful to recover the transition probability also in more general contexts than the BS model: result (2.38) – first shown by Breeden and Litzenberger (1978) – hinges on the particular form of the call payoff function $\psi(S_T) = (S_T - K)^+$, only, and is thus applicable in more general circumstances, irrespective of the particular distributional assumptions on the underlying asset price process. It derives its importance from the results of Sect. 2.3: if one knew this density, either by believing in the BS model or by obtaining an empirical estimate of it, any path independent contingent claim could be priced by simply integrating the payoff function over this density. The state price density is also useful for trading strategies, which try to exploit systematic deviances between the risk neutral and the historical properties of the underlying stock price time series, Aït-Sahalia et al. (2001b) and Blaskowitz et al. (2004). And it shall play an important role to derive local volatility, see Chap. 3 and in particular Sect. 3.3.

For this reason, the statistical literature has developed a whole battery of methods for estimating state price densities from observed option prices, see e.g. Jackwerth (1999) or Weinberg (2001) for reviews. The extraction of the state price density can be achieved for instance via parametric specifications of the density, or – in a discrete way – via implied trees, Sect. 3.10.1. This is more deeply discussed in Härdle and Zheng (2002). Recent advances by Härdle and Yatchew (2003) and Hlávka (2003) allow to estimate the state price density via non- and semiparametric procedures.

Finally, there is an identity which is useful for a lot of manipulations of the BS formula, see for instance Equation (2.34):

$$e^{-r\tau} K \varphi(d_2) = e^{-\delta\tau} S_t \varphi(d_1) \,, \tag{2.42}$$

which we state for completeness.

2.5 The IV Smile

It is obvious that the BS formula is derived under assumptions that are unlikely to be met in practice: frictionless markets, the ability to hedge continuously without transaction costs, asset prices without jumps, but independent Gaussian increments, and last but not least, a constant volatility function. Due to the simplicity of the model, any deviation from these assumptions is empirically summarized in one single parameter or object: the IV smile and the IVS.

The only unknown parameter in the BS pricing formula (2.23) is the volatility. Given observed market prices $\widetilde{C}_t$, it is therefore natural to define an *implicit* or *implied volatility* (IV), first introduced by Latané and Rendelman (1976):

$$\widehat{\sigma}: \quad C^{BS}(S_t, t, K, T, \widehat{\sigma}) - \widetilde{C}_t = 0 \,. \tag{2.43}$$

IV is the empirically determined parameter that 'makes the BS formula fit market prices of options'. Since the BS price is monotone in σ, as can be inferred from the positiveness of the call vega (2.30), there exists a unique solution $\widehat{\sigma} > 0$. Numerically, $\widehat{\sigma}$ can be found e.g. by a Newton-Raphson algorithm as discussed in Manaster and Koehler (1982). Finally, by the put-call-parity (2.26), put and call IV are equal.

In the derivation of the BS model it is presumed that the diffusion coefficient of the Brownian motion is a constant. IV $\widehat{\sigma}$, however, displays a pronounced curvature across option strikes K and, albeit to a lesser extent, across different expiry days T. Thus IV is in fact a mapping from time, strike prices and expiry days to $\mathbb{R}^+$:

$$\widehat{\sigma}: (t, K, T) \to \widehat{\sigma}_t(K, T) \,. \tag{2.44}$$

This mapping is called the *implied volatility surface* (IVS).

20   2 The Implied Volatility Surface

Often it is not convenient to work in absolute variables as expiry dates T and strikes K. Rather one prefers relative variables, since the analysis becomes independent of expiry effects and the movements of the underlying. Moreover, the options with strikes close to the spot price of the underlying asset are traded with high liquidity. As a new scale, one typically employs time to maturity $\tau \stackrel{\text{def}}{=} T - t$ and *moneyness*. In this work, we will predominantly use the following forward (or futures) moneyness definition:

$$\kappa_f \stackrel{\text{def}}{=} K/F_t , \qquad (2.45)$$

where $F_t = e^{(r-\delta)(T-t)} S_t$ denotes the (fair) futures or forward price at time t, Hull (2002). A stock price moneyness can be defined by:

$$\kappa \stackrel{\text{def}}{=} K/S_t . \qquad (2.46)$$

Forward moneyness is a natural choice of the moneyness scale, when one works with European style option data. European options can only be exercised at expiry. From this point of view, one incorporates the risk neutral drift in the moneyness measure, which is taken into account by dividing by the futures price.

We say that an option is at-the-money (ATM) when $\kappa \approx 1$. A call option is called out-of-the-money, OTM, (in-the-money, ITM), if $\kappa > 1$ ($\kappa < 1$) with the reverse applying to puts. Sometimes the literature also works in units of log-moneyness: $\ln(K/F_t)$ or $\ln(K/S_t)$. Given a quantity in one moneyness definition, it is often not difficult to switch between the different scales.

A typical picture of the IV smile is presented in Figs. 2.5 and 2.7. IV observations appear as black dots. The IV data, which are the basis for all empirical parts of this study, are obtained from prices of DAX index options traded at the EUREX in Frankfurt am Main. The original raw data was provided from the Deutsche Börse AG, Frankfurt. It has undergone considerable refinement and is stored in the financial data base MD*base located at the Center for Applied Statistics and Economics (CASE) at the Humboldt-Universität zu Berlin. A detailed description of the data and the preparation scheme is given in Appendix A. The option data is contract based data: each price observation belongs to *actual trades*, i.e. we do not work with price quotations or settlement data. Due to the nature of transaction based data, the data set may contain noise, potential misprints and other errors. This is also seen in Figs. 2.5 and 2.7 with the two single observations traded at an IV of 21% in the lower left of the smile function.

Figure 2.5 shows a downward-sloping smile across strikes for the 45 days to expiry contract as observed on 20000502. Obviously, OTM puts and ITM calls are traded at higher prices than the corresponding ATM options. Since the contracts are highly standardized on organized markets, IV observations are only available for a small subset of strikes. Consequently, observations are concentrated at these strikes.

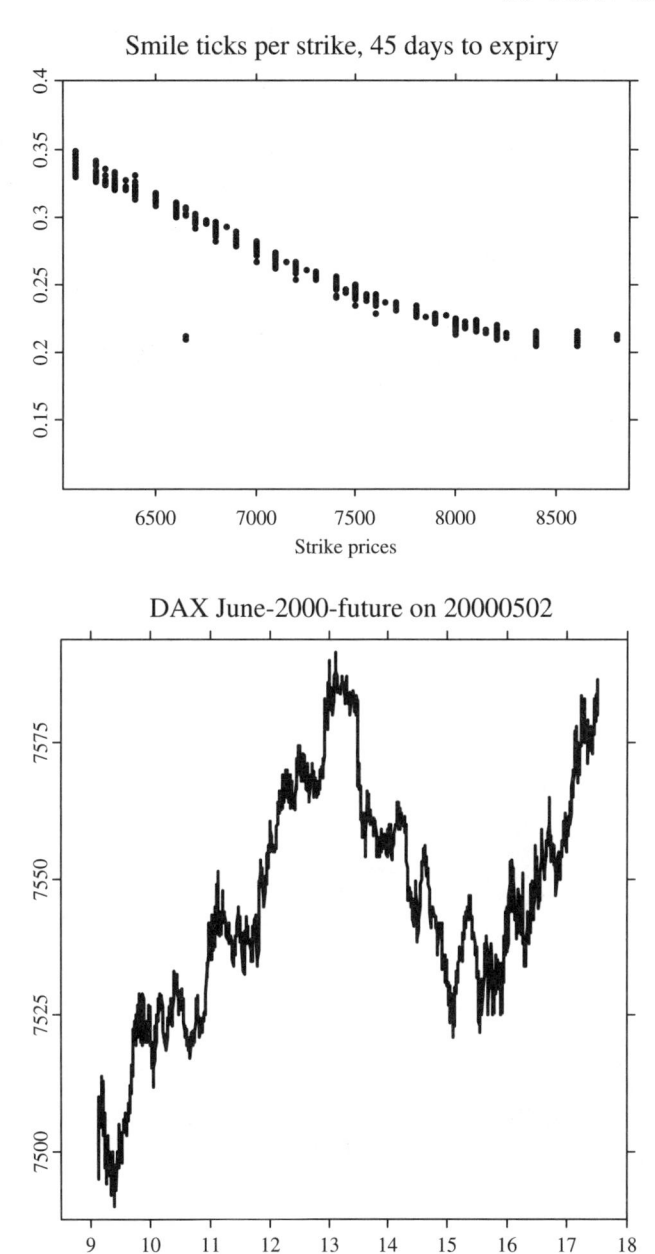

Fig. 2.5. *Top panel*: DAX option IV smile for 45 days to expiry on 20000502 plotted per strike. IV observations are displayed as black dots. *Bottom panel*: DAX futures contract, June 2000 contract, between 8:00 a.m. and 5:30 p.m. on 20000502
 SCMivanalysis.xpl

22   2 The Implied Volatility Surface

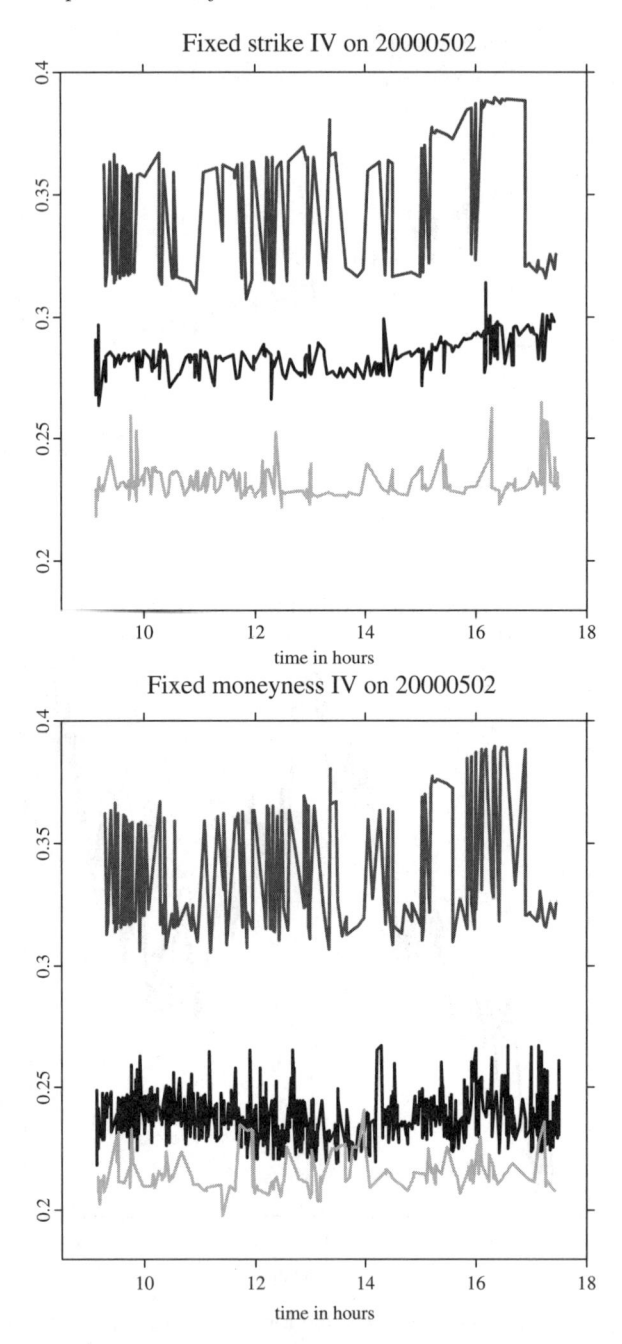

Fig. 2.6. *Top panel*: IV smile for 45 days to expiry plotted per strikes 6400 (*top*), 7000 (*middle*), and 7500 (*bottom*), between 8:00 a.m. and 5:30 p.m. on 20000502. *Bottom panel*: same IV smile plotted per (*forward*) moneyness 0.85 (*top*), 1.00 (*middle*), and 1.05 (*bottom*). `SCMivanalysis.xpl`

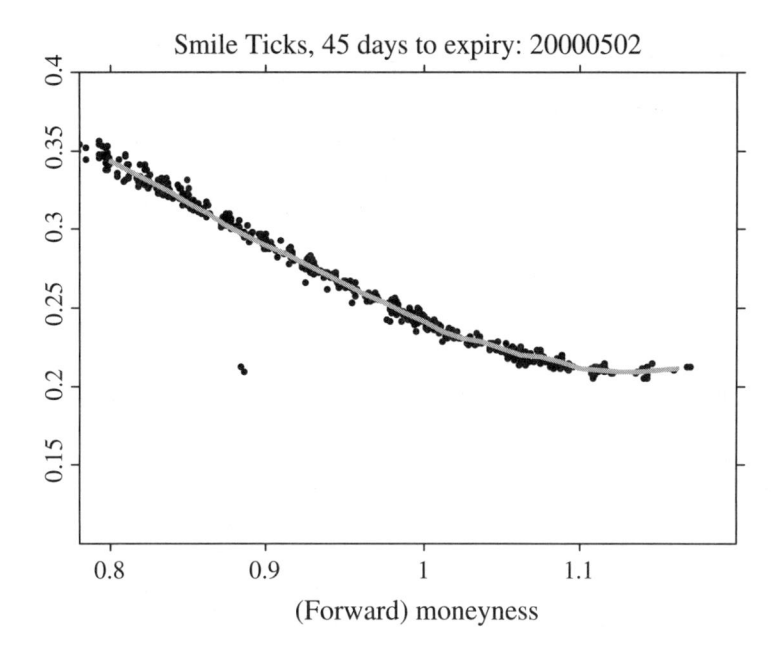

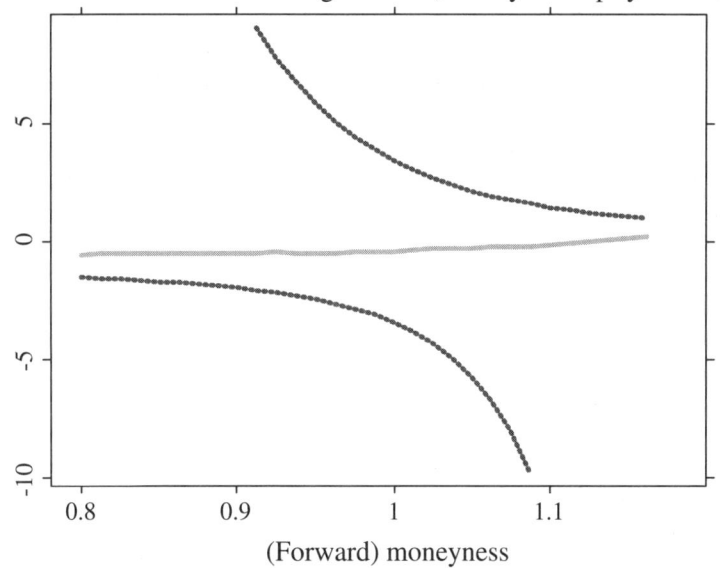

Fig. 2.7. *Upper panel*: IV smile for 45 days to expiry on 20000502. IV observations are displayed as *black* dots; the smile estimate is obtained from a local linear estimator with localized bandwidths. *Lower panel*: first order derivative obtained from a local linear estimator with localized bandwidths (*solid line*). No-arbitrage bounds (2.53) on the smile (*dashed*) SCMsmile.xpl

In the lower panel of Fig. 2.5, we added the intraday movements of the futures contract (expiry June 2000). Given the observation that the futures contract gains approximately 1% during the day, one may ask whether the dispersion of IV observations for a fixed strike is due to intraday movements of IV. In the top panel of Fig. 2.6, we present the intraday movements of IV at the fixed strikes 6400, 7000 and 7500. No particular directional moves of IV are evident. Rather – this is especially pronounced for the 6400-strike contract – IV jumps up and down between two distinct levels: this is the bid-ask bounce. During the day, the bid-ask spread seems to widen beginning from 3:00 p.m. Note that this coincides with a strong increase of the futures price in this part of the day. This contract, which is already in the OTM put region, is floating further away from the ATM region. The other contracts, closer to ATM, exhibit a much less pronounced jump behavior due to the bid-ask prices of the options.

In Fig. 2.7, the very same IV data are plotted against (forward) moneyness as defined in (2.45). As a proxy, we divide the strike of each option by the futures price which is closest in time within an interval of five minutes. It should be remarked that due to the daily settlement of futures contracts, futures prices and forward prices are not equal, when interest rates are stochastic. However, for the time to maturities we consider throughout this work (up to half a year), we believe this difference to be negligible, see Hull (2002, p. 51–52) for a more detailed discussion and further references to this topic. As is seen in Fig. 2.7, the overall shape of the smile function is not altered, but the data appear smeared across moneyness, which is due to the intraday fluctuations of the futures price. The lower panel of Fig. 2.6 gives the intraday movements for fixed moneyness in the neighborhood of $\kappa_f = 0.85, 1.00, 1.10$ (maximum distance is ± 0.02). It is seen that the turnover for ITM put (OTM call) options is very thin compared with OTM puts. Most trading activity is taking place ATM.

Comparing Fig. 2.5 with Fig. 2.7 exhibits a nice feature of the moneyness data. Plotting the smile against moneyness not only makes the smile independent from large moves in the underlying asset in the view of months and years. To some extent, it acts as a 'smoothing' device. This facilitates the aggregation of intraday data to daily samples as we do throughout this work. Especially, from the perspective of curve estimation, which is the topic of Chaps. 4 and 5, moneyness data are better tractable and more convenient.

Finally, let us present the entire IVS in Fig. 2.8. The IV smiles appear as black rows, which we shall call *strings*. The strings belong to different maturities of the option contracts. Similarly to the discrete set of strikes, only a very small number of maturities, here five, are actively traded at the same time. Also, one can discern that not all maturity strings are of comparable size: the third one is much shorter than the others. Obviously the IVS has a *degenerated design*. This poses several challenges to the modeling task which will be addressed in Sect. 5.4.

2.5 The IV Smile

IV ticks and IVS: 20000502

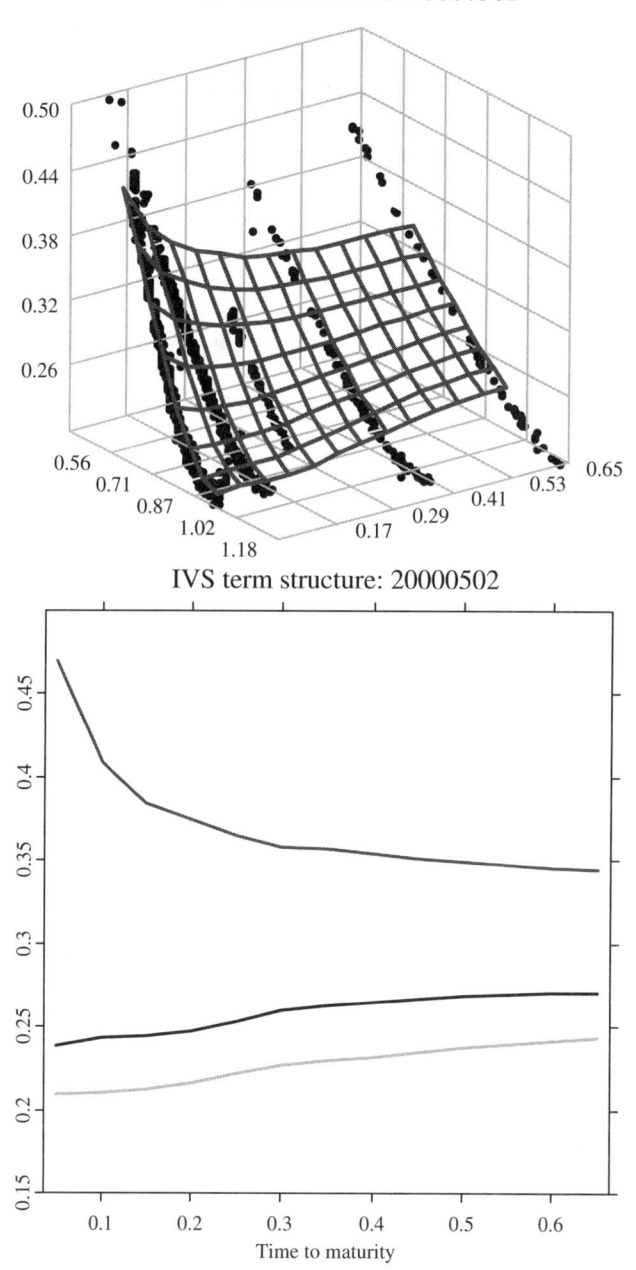

IVS term structure: 20000502

Fig. 2.8. *Top panel*: DAX option IVS on 20000502. IV observations are displayed as black dots; the surface estimate is obtained from a local quadratic estimator with localized bandwidths. *Bottom panel*: term structure of the IVS. $\kappa_f = 0.75$ top line, $\kappa_f = 1$, i.e. ATM, middle line, $\kappa_f = 1.1$, bottom line SCMivsts.xpl

As a general pattern, it is seen from Fig. 2.8 that the smile curve flattens out with longer time to maturity. The lower panel shows the term structure for various slices in the IVS for moneyness $\kappa_f = 0.75$ top line, $\kappa_f = 1$, i.e. ATM, middle line, $\kappa_f = 1.1$, bottom line. There is a slightly increasing slope for ATM IV and OTM call (ITM put) IV, while OTM put (ITM call) IV displays a decreasing term structure. This is due to the more shallow smiles for the long-term maturities.

The most fundamental conclusion of this section is that OTM puts and ITM calls are traded at higher prices than the corresponding ATM options. Obviously, the BS model does *not* properly capture the probability of large downward movements of the underlying asset price. To arrive at an explanation, one needs to relax the assumptions of the BS model. This literature is summarized in Sect. 2.11.

The pivotal question following this conclusion is: *what does the IVS imply for practice?* Two points shall be raised here:

The good, or perhaps lucky news is that the ambiguity of the model is sublimated in one single entity, the IV. This allows traders to think themselves as making a market for volatility rather than for specific equity contracts: hence, it is common practice to quote options in terms of IV. The BS formula is only employed as a simple and convenient mapping to assign to each option on the same underlying a strike-dependent (and a maturity-dependent) IV. For this purpose it is not necessary to believe in the BS model. It simply acts as a computational tool insuring a common language among traders.

The bad news is that for each K and each T across the IVS a *different BS model applies*. This causes difficulties in managing option books, as shall be discussed in Sect. 2.9. The reason is that for hedging purposes it may not be a good idea to evaluate the delta of the option using its own 'quoted IV'. Also for pricing exotic options, the presence of the IVS poses challenges, especially for volatility-sensitive exotics, such as barrier options. This issue is addressed in Chap. 3.

Often, IV is interpreted as the market's expectation of *average* volatility through the life time of the option. At first glance this notion seems sensible, since if the market has a consensus about future volatility, it will be reflected in IV. Unfortunately, from a theoretical point of view, this notion can only be validated for a very limited class of models, as shall be demonstrated in Sect. 2.8. Furthermore, option markets are also driven by supply and demand. If market participants seek for some reason protection against a down-swing in the market, this will drive up put prices. Eventually, since the put price (as the call price) is positively monotonous in volatility, this will be reflected in higher IVS levels for OTM puts. Thus, this notion should be treated with caution.

2.6 Static Properties of the Smile Function

2.6.1 Bounds on the Slope

From the general fact that (European) call prices are monotonically decreasing and puts are monotonically increasing functions of strike prices, compare (2.33), it is possible to obtain broad no-arbitrage bounds on the slope of the smile, Lee (2002). If $K_1 < K_2$ for any expiry date T, we have

$$C_t(K_1, T) \geq C_t(K_2, T) \quad , \quad P_t(K_1, T) \leq P_t(K_2, T) . \tag{2.47}$$

Due to an observation by Gatheral (1999), this can be improved to:

$$C_t(K_1, T) \geq C_t(K_2, T) \quad , \quad \frac{P_t(K_1, T)}{K_1} \leq \frac{P_t(K_2, T)}{K_2} . \tag{2.48}$$

Assuming the explicit dependence of volatility on strikes, we obtain by differentiating:

$$\frac{\partial C_t}{\partial K} = \frac{\partial C_t^{BS}}{\partial K} + \frac{\partial C_t^{BS}}{\partial \widehat{\sigma}} \frac{\partial \widehat{\sigma}}{\partial K} \leq 0 , \tag{2.49}$$

which implies

$$\frac{\partial \widehat{\sigma}}{\partial K} \leq -\frac{\partial C_t^{BS}/\partial K}{\partial C_t^{BS}/\partial \widehat{\sigma}} . \tag{2.50}$$

Differentiating P_t^{BS}/K with respect to K, yields for the lower bound:

$$\frac{\partial \widehat{\sigma}}{\partial K} \geq \frac{P_t^{BS}/K - \partial P_t^{BS}/\partial K}{\partial P_t^{BS}/\partial \widehat{\sigma}} . \tag{2.51}$$

Finally, insert the analytical expressions of the option derivatives and the put price and make use of relationship (2.42). This shows:

$$-\frac{\Phi(-d_1)}{\sqrt{\tau} K \varphi(d_1)} \leq \frac{\partial \widehat{\sigma}}{\partial K} \leq \frac{\Phi(d_2)}{\sqrt{\tau} K \varphi(d_2)} . \tag{2.52}$$

Using $F_t = e^{(r-\delta)(T-t)} S_t$ and $\kappa_f = K/F_t$, this can be expressed in terms of our forward moneyness measure as:

$$-\frac{\Phi(-d_1)}{\sqrt{\tau} \kappa_f \varphi(d_1)} \leq \frac{\partial \widehat{\sigma}}{\partial \kappa_f} \leq \frac{\Phi(d_2)}{\sqrt{\tau} \kappa_f \varphi(d_2)} , \tag{2.53}$$

since $\partial \widehat{\sigma}/\partial K = \partial \widehat{\sigma}/\partial \kappa_f F_t^{-1}$.

The bounds are displayed in the lower panel of Fig. 2.7 together with the estimated first order derivative. It is seen that the bounds are very broad given the estimated slope of the smile function. Without the refinement (2.48), the lower bound is $-\{1 - \Phi(d_2)\}/\sqrt{\tau} \kappa_f \varphi(d_2)$, only.

2.6.2 Large and Small Strike Behavior

Lee (2003) derived remarkable results for the large and small strike behavior of the smile function. Define $x \stackrel{\text{def}}{=} \ln \kappa_f = K/F_t$, where the futures prices remains fixed in this section. He shows that

$$\widehat{\sigma}_t(x,T) < \sqrt{\frac{2|x|}{T}} \qquad (2.54)$$

for some sufficiently large $|x| > x^*$ (see also Zhu and Avellaneda (1998) who derive this bound in a less general setting). He proceeds in showing that there is a precise one-to-one correspondence between the asymptotic behavior of the smile function and the number of finite moments of the distribution of the underlying S_T and its inverse, S_T^{-1}.

To understand the first result consider the large strike case. Due to monotonicity of the BS formula in volatility, it is equivalent to (2.54) to show for some $x > x^*$ that

$$C_t^{BS}(x,T,\widehat{\sigma}_t(x,T)) < C_t^{BS}(x,T,\sqrt{2|x|/T}) \,. \qquad (2.55)$$

For the left-hand side one sees for any call price function that

$$\lim_{x \uparrow \infty} C_t(x,T) = \lim_{K \uparrow \infty} e^{-r\tau} \mathsf{E}^Q (S_T - K)^+ = 0 \,, \qquad (2.56)$$

since by $\mathsf{E}^Q S_T < \infty$ we can interchange the limit and the expectation by the dominated convergence theorem. For the right-hand side one obtains

$$\lim_{x \uparrow \infty} C_t^{BS}(x,T,\sqrt{2|x|/T}) = e^{-r\tau} F_t \left\{ \Phi(0) - \lim_{x \uparrow \infty} e^x \Phi(-\sqrt{2|x|}) \right\}$$
$$= e^{-r\tau} F_t / 2 \,, \qquad (2.57)$$

by applying L'Hôpital's rule. A similar approach involving puts proves the small strike case, Lee (2003, Lemma 3.3).

The second result relates a coefficient that can replace the '2' in (2.54) with the number of finite moments in the underlying distribution of S_T and S_T^{-1}. Define

$$\widetilde{p} \stackrel{\text{def}}{=} \sup\{p : \mathsf{E} S_T^{1+p} < \infty\} \,, \qquad (2.58)$$

$$\widetilde{q} \stackrel{\text{def}}{=} \sup\{q : \mathsf{E} S_T^{-q} < \infty\} \,, \qquad (2.59)$$

and

$$\beta_R \stackrel{\text{def}}{=} \limsup_{x \uparrow \infty} \frac{\widehat{\sigma}^2(x,T)}{|x|/T} \,, \qquad (2.60)$$

$$\beta_L \stackrel{\text{def}}{=} \limsup_{x \downarrow -\infty} \frac{\widehat{\sigma}^2(x,T)}{|x|/T} \,. \qquad (2.61)$$

2.6 Static Properties of the Smile Function

The coefficients β_R, β_L can be interpreted as the slope coefficients of the asymptotes of the implied variance function.

Lee (2003, Theorem 3.2 and 3.4) shows that $\beta_R, \beta_L \in [0, 2]$. Moreover, he proves that

$$\widetilde{p} = \frac{1}{2\beta_R} + \frac{\beta_R}{8} - \frac{1}{2}, \qquad (2.62)$$

$$\widetilde{q} = \frac{1}{2\beta_L} + \frac{\beta_L}{8} - \frac{1}{2}, \qquad (2.63)$$

where $1/0 \stackrel{\text{def}}{=} \infty$, and

$$\beta_R = 2 - 4\left(\sqrt{\widetilde{p}^2 + \widetilde{p}} - \widetilde{p}\right), \qquad (2.64)$$

$$\beta_L = 2 - 4\left(\sqrt{\widetilde{q}^2 + \widetilde{q}} - \widetilde{q}\right), \qquad (2.65)$$

where β_R, β_L are read as zero, if $\widetilde{p}, \widetilde{q}$ are infinity.

The intuition of these results is as follows: the IV smile must carry the same information as the underlying risk neutral transition density. For instance, in the empirical literature it is well known that certain shapes of the state price density determine the shape of the smile function. Also the asymptotic behavior of the smile is shaped by the tail behavior of the risk neutral transition density and vice versa. The tail decay of the risk neutral transition density, however, determines the number of finite moments in the distribution. This context is made rigorous by employing two fundamental results: on the one hand, options are bounded by moments, Broadie et al. (1998), and on the other hand, moments, which can be interpreted as exotic options with power payoffs, are bounded by mixtures of a strike continuum of plain vanilla options, Carr and Madan (1998).

The results by Lee (2003) have implications for the extrapolation of the IVS, for instance in the context of local volatility models, see in Sect. 3.10.3, pp. 87. In a meaningful pricing algorithm it is often necessary to extrapolate the IVS beyond the values at which options are typically observed. The choice of the extrapolation is an intricate task since the prices of exotic options depend significantly on the specific extrapolation. For instance, assuming that at large and small strikes the IVS flattens out to a constant may produce prices of exotic options that are far different from those obtained in a model that allows the IVS to moderately increase at large and small strikes. The results now provide guidelines for this extrapolation: the smile should not grow faster than $\sqrt{|x|}$. Furthermore, it should not grow slower than $\sqrt{|x|}$ *unless* one assumes that S_T has finite moments of all orders. Finally, the moment formula allows to determine the number of finite moments that one immediately implies in the choice of the multiplicative factor in the growth term that extrapolates the smile.

2.7 General Regularities of the IVS

2.7.1 Static Stylized Facts

Despite its daily fluctuations, the IVS exhibits a number of empirical regularities, both from a static and a dynamic perspective. Here, they shall be summarized with respect to the DAX index options traded at the German option market, Deutsche Börse AG, Frankfurt, see Appendix A for details concerning the data. Typically these stylized facts are observed for any equity index market. Markets with other underlying assets display similar features. For a compendium on single stocks, interest rate and foreign exchange markets see Rebonato (1999) or Tompkins (1999).

1. For short time to maturities the smile is very pronounced, while the smile becomes more and more shallow for longer time to maturities: the IVS flattens out, Fig. 2.8. Figure 2.9 displays the mean IVS 1996. Since this is computed from smoothed IVS data and on a relatively small grid, this effect is less apparent than in Fig. 2.8. For more pictures of this kind, see Fengler (2002).
2. The smile function achieves its minimum in the neighborhood of ATM to near OTM call options, Fig. 2.7 and Fig. 2.9. The term structure is increasing, but may also display a humped profile, especially in periods of market turmoil as during the Asian crisis in 1997, Fig. 2.11.

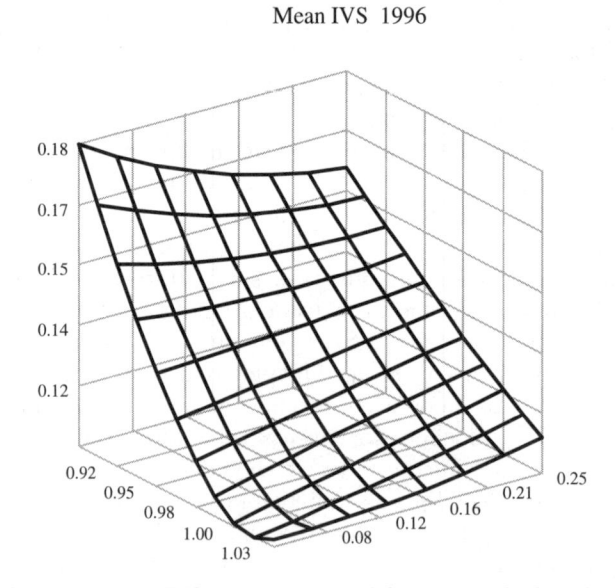

Fig. 2.9. Mean IVS 1996, computed from smoothed surfaces

2.7 General Regularities of the IVS 31

Standard Deviation of IVS 1996

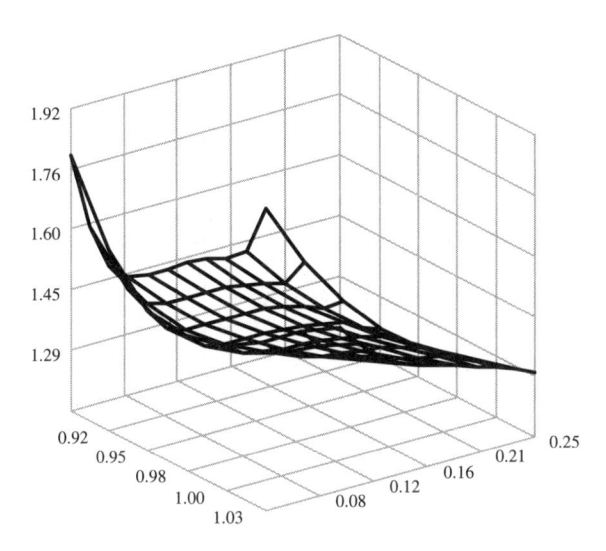

Fig. 2.10. Standard deviation of the IVS 1996, computed from smoothed surfaces

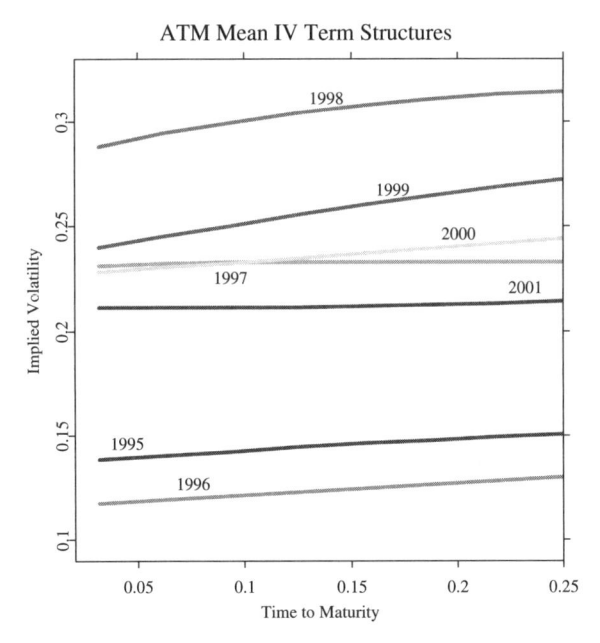

ATM Mean IV Term Structures

Fig. 2.11. ATM mean IV term structures between 1995 and 2001, computed from smoothed surfaces

3. OTM put regions display higher levels of IV than OTM call regions, lower panel of Fig. 2.8 and Fig. 2.9. However, this has not always been the case: a more or less symmetric smile became strongly asymmetric (a 'sneer', or 'smirk') and considerably more pronounced after the 1987 crash. It is widely argued that this is due to the investors' increased awareness of market down-swings since this period, Rubinstein (1994).
4. The volatility of IV is biggest for short maturity options and monotonically declining with time to maturity, Fig. 2.10 and Fig. 2.12, also Fengler (2002).
5. Returns of the underlying asset and returns of IV are negatively correlated, indicating a leverage effect, Black (1976). For the entire data 1995 to May 2001 we find a correlation between ATM IV (three months) and DAX returns of $\rho = -0.32$. This point is further discussed in the context of the principal component analysis in Sect. 5.2.5.
6. IV appears to be mean-reverting, Cont and da Fonseca (2002). For ATM IV (three months) we find a mean reversion of approximately 60 days, see also Sect. 5.2.5 and Table 5.4.
7. Shocks across the IVS are highly correlated. Thus, IVS dynamics can be decomposed into a small number of driving factors, Chap. 5.

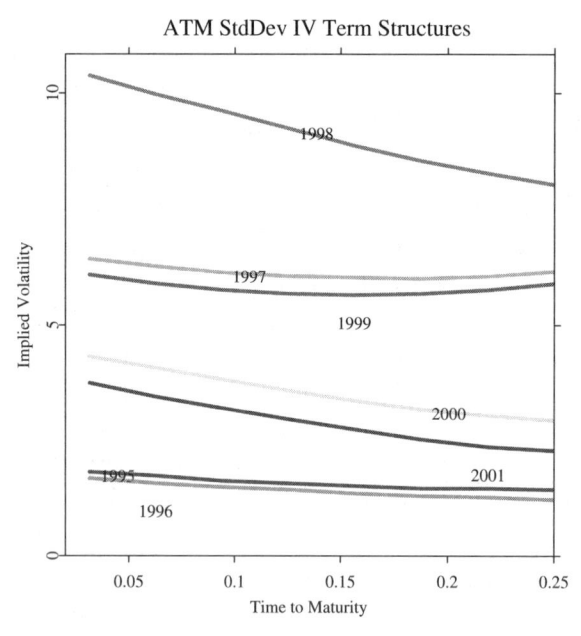

Fig. 2.12. ATM standard deviation of IV term structures between 1995 and 2001, computed from smoothed surfaces

2.7.2 DAX Index IV between 1995 and 2001

An overview on the three-month ATM IV time series between 1995 and May 2001 is given in Fig. 2.13. In Fig. 2.14, the same time series together with the rescaled DAX is shown. At the beginning of 1995, the DAX index was at around 2100 points and increased moderately till the beginning of 1997. During this time ATM IV was below 20% and gradually fell till 1997 towards 14%. Beginning from the end of 1996 the DAX commenced a steady and smooth increase till mid 1998, which was shortly interrupted by the Asian crisis in the second half of the year 1997. The entire ascent of the DAX was accompanied by steadily increasing IV levels rising as high as 35%, when the DAX fell sharply at the peak of the market turmoil. From then, IVs gradually declined, but remained – also relatively volatile – at historically high levels, while the index rose again. Between mid of July and beginning of October 1998 the DAX dropped again about 2000 points, followed by a sharp increase of IVs with peaks up to 50%. During the recovery of the index between 1999 and 2000, IVs returned to the levels recorded before the late 1998 increase. Although increasingly volatile, they remained at these levels, when the DAX began its gradual decline from the post war peak of 8000 points in March 2000.

The annual standard deviation of IV can be inferred from Fig. 2.12. As already observed short run volatilities are more subject to daily variation than

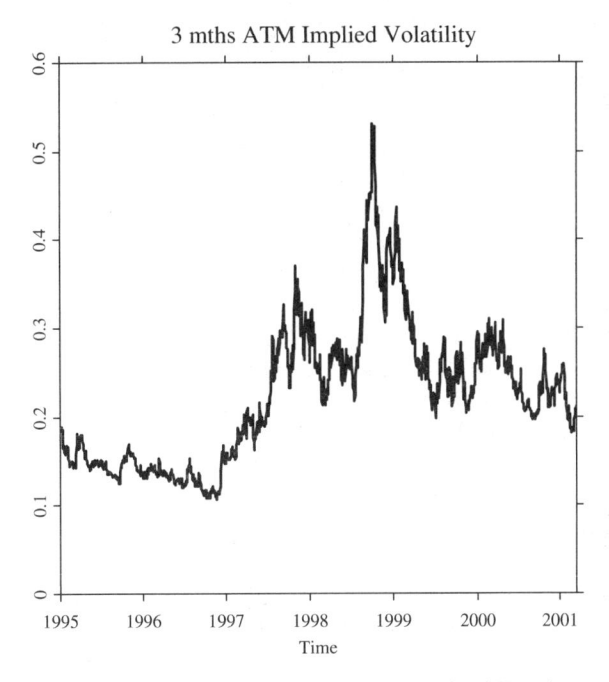

Fig. 2.13. The three-months ATM IV levels of DAX index options

ATM IV and DAX Index

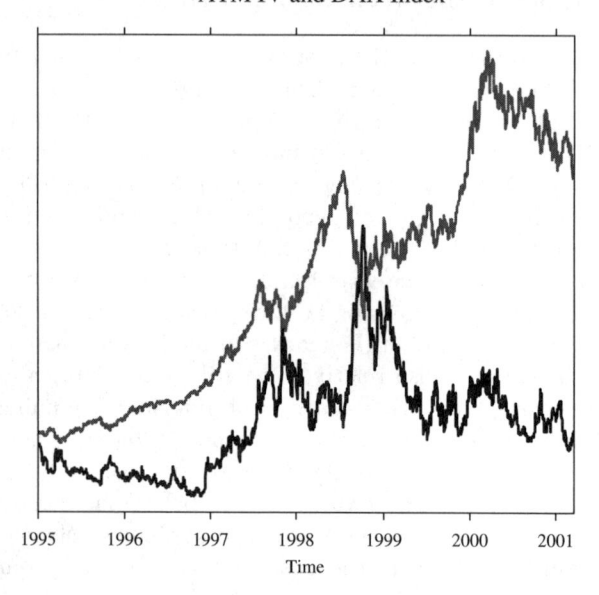

Fig. 2.14. German DAX $\times 10^{-4}$ (*upper line*) and three-months ATM IV levels (*lower line*), also given in Fig. 2.12

the long term volatilities, which is reflected in the downward sloping functions. The year 1998 was the period of the highest volatility, followed by the years 1997 and 1999.

From this description, the only obvious regularity between the time series patterns of the underlying asset, the DAX index, and ATM IV is that times of market crises lead to sharply increasing levels of the IVS. This is likely to be due to the increased demand for put options. Otherwise no clear-cut relation between the time series patterns emerges: levels of the IVS may be constant, downward or upward-trending independently from the index. Some authors, like Derman (1999) and Alexander (2001b), have suggested to distinguish different market regimes. In this interpretation, the IVS acts as an *indicator of market sentiment*, i.e. as an additional financial variable describing the current state of the market. This view explains the increased interest in accurate modeling techniques of the IVS.

2.8 Relaxing the Constant Volatility Case

The fact that IV is counterfactual to the BS model has spurred a large number of alternative pricing models. The easiest way for more flexibility is to allow the coefficients of the SDE, which describes the stock price evolution, to be deterministic functions in the asset price and time. This preserves the complete market setting, Sect. 2.8.1. A second important class of models specifies volatility as an additional stochastic process. Since volatility is not a tradable asset, this implies that the market is incomplete. A short review is given in Sect. 2.8.2.

2.8.1 Deterministic Volatility

Allowing for coefficients of the SDE of the stock price evolution that are deterministic functions in the asset price and time, leads to

$$\frac{dS_t}{S_t} = \mu(S_t, t)\,dt + \sigma(S_t, t)\,dW_t \,, \tag{2.66}$$

where $\mu, \sigma : \mathbb{R} \times [0, T^*] \to \mathbb{R}$ are deterministic functions. For the existence of a unique strong solution, the functions must satisfy a global Lipschitz and a linear growth condition, see appendix Chap. B.

For pricing derivatives, one may proceed as in Sect. 2.1. This leads to the *generalized BS PDE* for the derivative price:

$$0 = \frac{\partial H}{\partial t} + (r - \delta)S\frac{\partial H}{\partial S} + \frac{1}{2}\sigma^2(S, t)S^2\frac{\partial^2 H}{\partial S^2} - rH \,. \tag{2.67}$$

As before the delta hedge ratio is given by the first order derivative of the solution to (2.67) with respect to S_t.

For plain vanilla options, closed-form solutions can be derived for some particular specifications of volatility. Generally, this can be achieved via change of variable techniques, e.g. Bluman (1980) and Harper (1994). For the special case, when volatility is only time-dependent, i.e. $\sigma(S_t, t) \stackrel{\text{def}}{=} \sigma(t)$, one can use the following arguments to solve (2.67), Wilmott (2001a, Chap. 8):

One introduces the new variables

$$\overline{S} \stackrel{\text{def}}{=} S e^{(r-\delta)(T-t)} \,, \tag{2.68}$$

$$\overline{t} \stackrel{\text{def}}{=} f(t) \,, \tag{2.69}$$

$$\overline{H}(\overline{S}, \overline{t}) \stackrel{\text{def}}{=} H(S, t)\,e^{r(T-t)} \,, \tag{2.70}$$

where f is some smooth function. Expressing the PDE (2.67) in terms of the new variables (2.68) to (2.70), yields

$$\frac{\partial \overline{H}}{\partial \overline{t}}\frac{\partial \overline{t}}{\partial t} = \frac{1}{2}\sigma^2(t)\overline{S}^2\frac{\partial^2 \overline{H}}{\partial \overline{S}^2} \,. \tag{2.71}$$

If we choose

$$f(t) \stackrel{\text{def}}{=} \int_t^T \sigma^2(s)\,ds \,, \tag{2.72}$$

Equation (2.71) reduces to

$$\frac{\partial \overline{H}}{\partial \overline{t}} = \frac{1}{2}\overline{S}^2 \frac{\partial^2 \overline{H}}{\partial \overline{S}^2} , \qquad (2.73)$$

which is *independent* from time in its coefficients. Also, the boundary condition for a (European) call $\overline{H}(\overline{S}_T, \overline{T}) = H(S_T, T) = (S_T - K)^+$ (or a European put) stays the *same* after these manipulations. Consequently, in denoting by $\overline{H}(\overline{S}, \overline{t})$ the solution to (2.73), we can rewrite this in the original variables as

$$H(S,t) = e^{-r(T-t)} \overline{H}\{Se^{(r-\delta)(T-t)}, f(t)\} . \qquad (2.74)$$

Now denote by H^{BS} the BS solution with constant volatility σ. It can be written in the form

$$H^{BS} = e^{-r(T-t)} \overline{H}^{BS}\{Se^{(r-\delta)(T-t)}, \overline{\sigma}^2(T-t)\} , \qquad (2.75)$$

for some function $\overline{H}^{BS}$. By comparison of (2.74) and (2.75), it is seen that the constant and the time-dependent coefficient case have the same solutions if we put

$$\overline{\sigma}^2 \stackrel{\text{def}}{=} \frac{1}{T-t} f(t) = \frac{1}{T-t} \int_t^T \sigma^2(s)\, ds . \qquad (2.76)$$

Hence, in the case of a European call option we have

$$C(S_t, t) = C^{BS}\left(S_t, t, K, T, \sqrt{\overline{\sigma}^2}\right) . \qquad (2.77)$$

This is the common BS formula with the volatility parameter σ *replaced by an average volatility up to expiry*. It follows by the definition of IV in (2.43) that

$$\widehat{\sigma} = \sqrt{\overline{\sigma}^2} . \qquad (2.78)$$

Therefore, this result provides a justification for interpreting IV as the *average* volatility over the option's life time from t to T, Sect. 2.5.

The time-dependent volatility generates a term structure of the IVS, only, *not* a smile. To obtain a smile, volatility must depend on S_t as well. This is achieved in the more general *local volatility models*. Actually, the model with time-dependent volatility is just a special case of local volatility models. Consequently, the further discussion of these models is delayed to Chap. 3.

2.8.2 Stochastic Volatility

In stochastic volatility models, additionally to the Brownian motion which drives the asset price a second stochastic process is introduced that governs the volatility dynamics. A typical model set-up can be given by:

$$\frac{dS_t}{S_t} = \mu\, dt + \sigma(t, Y_t)\, dW_t^{(0)} , \qquad (2.79)$$

$$\sigma(t, Y_t) = f(Y_t) ,$$

$$dY_t = \alpha(Y_t, t)\, dt + \theta(Y_t, t)\, dW_t^{(1)} . \qquad (2.80)$$

2.8 Relaxing the Constant Volatility Case

The two Brownian motions $\left(W_t^{(0)}\right)_{0\leq t\leq T^*}$ and $\left(W_t^{(1)}\right)_{0\leq t\leq T^*}$ are defined on the probability space $(\Omega, \mathcal{F}, \mathrm{P})$, and let $(\mathcal{F}_t)_{0\leq t\leq T^*}$ be the P-augmented filtration generated by both Brownian motions. Again we suppose that the sufficient conditions are met, such that (2.79) and (2.80) have unique strong solutions, Chap. B. The function $f(y)$ chosen for positivity and analytical tractability: typical examples are: $f(y) = \sqrt{y}$, Hull and White (1987), $f(y) = e^y$, Stein and Stein (1991), or $f(y) = |y|$, Scott (1987).

Empirical analysis suggests a mean-reverting behavior of volatility. To capture this regularity, $(Y_t)_{0\leq t\leq T^*}$ is often assumed to be an Ornstein-Uhlenbeck process, which is defined as the solution to the SDE

$$dY_t = \alpha(\mu_y - Y_t)\,dt + \theta\,dW_t^{(1)}, \tag{2.81}$$

where $\alpha, \mu_y, \theta > 0$, Scott (1987) and Stein and Stein (1991). Here, α is the rate of mean reversion, pulling the levels of the process back to its long run mean μ_y. The solution of (2.81) is given by:

$$Y_t = \mu_y + (y_0 - \mu_y)e^{-\alpha t} + \theta\int_0^t e^{-\alpha(t-s)}\,dW_s^{(1)}, \tag{2.82}$$

where y_0 denotes a known starting value. The distribution of Y_t is

$$N\left(\mu_y + (y_0 - \mu_y)e^{-\alpha t}, \frac{\theta^2}{2\alpha}(1 - e^{-2\alpha t})\right). \tag{2.83}$$

The stationary distribution for $t \uparrow \infty$ is given by $N\left(\mu_y, \frac{\theta^2}{2\alpha}\right)$, which does not depend on y_0.

Alternatively, the literature considers a log-normal process (Hull and White; 1987), or the Cox-Ingersoll-Ross process, Heston (1993) and Ball and Roma (1994), or a combination of the constant elasticity of variance model with stochastic volatility, Hagan et al. (2002), the so called SABR model.

Usually, it is supposed that both Brownian motions are correlated, i.e.

$$\langle W^{(0)}, W^{(1)}\rangle_t = \rho t, \tag{2.84}$$

where $-1 \leq \rho \leq 1$ is the instantaneous correlation between the processes. The case $\rho < 0$ is associated with the *leverage* effect, Black (1976): volatility rises, when there is a negative shock in the market value of the firms, since this results in an increase in the debt-equity ratio. This pattern is also observed for IV processes, Sect. 2.7.

As has been stated earlier, the assumption of no-arbitrage is equivalent to the existence of an equivalent martingale measure. Unlike to Sect. 2.1, the market is not complete due to the additional source of risk. In general, there exists an entire set of equivalent martingale measures $\mathcal{Q}$. A martingale measure $Q \in \mathcal{Q}$ can be characterized by the Radon-Nikodým derivative using Girsanov's Theorem, Chap. B:

2 The Implied Volatility Surface

$$\frac{dQ}{dP} = \exp\left\{-\int_0^{T^*} \lambda_s^{(0)} W_s^{(0)} \, ds - \frac{1}{2}\int_0^{T^*} \left(\lambda_s^{(0)}\right)^2 ds \right.$$
$$\left. - \int_0^{T^*} \lambda_s^{(1)} W_s^{(1)} \, ds - \frac{1}{2}\int_0^{T^*} \left(\lambda_s^{(1)}\right)^2 ds \right\}, \quad (2.85)$$

where $\left(\lambda_t^{(0)}\right)_{t\geq 0}$ and $\left(\lambda_t^{(1)}\right)_{t\geq 0}$ are $(\mathcal{F}_t)_{0\leq t\leq T^*}$-adapted processes. Furthermore,

$$\overline{W}_t^{(0)} \stackrel{\text{def}}{=} W_t^{(0)} + \int_0^t \lambda_s^{(0)} \, ds \quad \text{and} \quad \overline{W}_t^{(1)} \stackrel{\text{def}}{=} W_t^{(1)} + \int_0^t \lambda_s^{(1)} \, ds, \quad (2.86)$$

are Brownian motions on the space $(\Omega, \mathcal{F}, Q)$ for all $t \in [0, T^*]$. If (and only if)

$$\lambda_t^{(0)} \stackrel{\text{def}}{=} \frac{\mu + \delta - r}{\sigma(t, Y_t)}, \quad (2.87)$$

the discounted price process $e^{-rt} S_t$ is a martingale. The process $\left(\lambda_t^{(1)}\right)_{t\geq 0}$ can be any adapted process satisfying the required integrability condition. In analogy to (2.87) it is called the *market price of volatility risk*. The measure Q depends on the choice of $\left(\lambda_t^{(1)}\right)_{t\geq 0}$. In some sense, one may think about the measure as being 'parameterized' by this process, we write therefore: Q^{λ_1}. Option prices are computed by exploiting the risk neutral pricing relationship:

$$H_t = \mathsf{E}^{Q^{\lambda_1}}\left\{e^{-rt}\psi(S_T)|\mathcal{F}_t\right\}. \quad (2.88)$$

Let's assume that $\left(\lambda_t^{(1)}\right)_{t\geq 0}$ is a function of Y_t, S_t and t only, i.e. a Markov process, and that $\rho \stackrel{\text{def}}{=} 0$. In this particular case, we can compute option prices by conditioning on the volatility path. By the law of iterated expectations, we have, e.g. for a call:

$$C(S_t, t) = \mathsf{E}^{Q^{\lambda_1}}\left[\mathsf{E}^{Q^{\lambda_1}}\left\{e^{-r(T-t)}(S_T - K)^+ | \mathcal{F}_t, \sigma(Y_s, s), t \leq s \leq T\right\} | \mathcal{F}_t\right]. \quad (2.89)$$

Since the inner expectation is the same as in the time-dependent volatility case, this reduces to

$$C(S_t, t) = \mathsf{E}^{Q^{\lambda_1}}\left\{C^{BS}\left(S_t, t, K, T, \sqrt{\overline{\sigma}^2}\right) | \mathcal{F}_t\right\}, \quad (2.90)$$

where now $\overline{\sigma}^2 \stackrel{\text{def}}{=} \frac{1}{T-t}\int_t^T \{f(Y_s)\}^2 \, ds$. As before we insert the root-mean-square time average over a particular trajectory of volatility into the BS formula. The call price is given by an average of prices over all possible volatility paths.

Due to its similarity to the case of deterministic volatility, one is tempted to interpret IV as an average volatility over the remaining life time of the option. However, in general we have

2.8 Relaxing the Constant Volatility Case

$$\mathsf{E}^{Q^{\lambda_1}}\left\{C^{BS}\left(\cdot,\sqrt{\overline{\sigma}^2}\right)|\mathcal{F}_t\right\} \neq C^{BS}\left(\cdot,\mathsf{E}^{Q^{\lambda_1}}\sqrt{\overline{\sigma}^2}|\mathcal{F}_t\right), \qquad (2.91)$$

and thus:

$$\widehat{\sigma} \neq \mathsf{E}^{Q^{\lambda_1}}\left(\sqrt{\overline{\sigma}^2}|\mathcal{F}_t\right), \qquad (2.92)$$

since $\overline{\sigma}^2$ is random and the call price a nonlinear function of volatility. For ATM strikes, where the volga is small and negative (recall our remark from page 15 with regard to Fig. 2.4), Jensen's inequality yields:

$$\widehat{\sigma} < \mathsf{E}^{Q^{\lambda_1}}\left(\sqrt{\overline{\sigma}^2}|\mathcal{F}_t\right), \qquad (2.93)$$

but $\widehat{\sigma} \approx \mathsf{E}^{Q^{\lambda_1}}\left(\sqrt{\overline{\sigma}^2}|\mathcal{F}_t\right)$ may be considered as a sufficiently good approximation.

Thus, it can still be justified to interpret IV as an average volatility over the remaining life time of the option. It should be borne in mind, however, that this interpretation is limited for ATM strikes and $\rho = 0$, only. For the case $\rho \neq 0$, a representation of this form depends strongly on the specification of the underlying volatility process. A generalization of (2.89) within the Hull and White (1987) model for any ρ has been obtained by Zhu and Avellaneda (1998), see also the discussion in Fouque et al. (2000).

Due to the incompleteness of the market, the construction of a riskless hedge portfolio as in Sect. 2.1 is not possible. One solution, referred to as *delta-sigma hedging*, is to introduce another option with a longer maturity into the market, whose price is given exogenously. This completes the market under some conditions, Bajeux and Rochet (1992). Given these three instruments, the stock, the bond and the additional option, a riskless hedge portfolio can be constructed that prices the option.

Another strategy is to assume that $\left(\lambda_t^{(1)}\right)_{t\geq 0} \stackrel{\text{def}}{=} 0$, i.e. volatility risk is unpriced. This is sensible if volatility risk can diversified away, or preferences are logarithmic, Pham and Touzi (1996). The measure Q^0 can also be interpreted as the closest measure to P in an relative entropy sense, Föllmer and Schweizer (1990). In the general case, one needs to resort to hedging strategies that have been developed within the incomplete markets literature: in *super-hedging* the contingent claim is 'super-replicated', i.e. a self-financing strategy with minimum initial costs is sought such that any future obligation from selling the contingent claim is covered, while in *quantile-hedging* one tries to cover this obligation only with a sufficiently high probability. Finally, one may consider trading strategies which are not necessarily self-financing, i.e. which allow for the additional transfer of wealth to the hedge portfolio. This is called *risk-minimizing hedging* originated by Föllmer and Sondermann (1986). See Föllmer and Schweizer (1990), Karatzas (1997) and Föllmer and Schied (2002) for a detailed mathematical treatment of these hedging approaches.

2.9 Challenges Arising from the Smile

2.9.1 Hedging and Risk Management

In the presence of the smile, a first obvious challenge is the computation of the relevant hedge ratios. At first glance, an answer may be to insert IV into the BS derivatives in order to compute the hedge ratios for some option position. This strategy is also called an 'IV compensated BS hedge'. However, one should be aware that this strategy can be erroneous, since IV is not necessarily equal to the *hedging volatility*. Analogously to IV, the hedging volatility, for instance for the delta, is defined by:

$$\widehat{\sigma}_h : \quad \frac{\partial C^{BS}}{\partial S}(S_t, t, K, T, \widehat{\sigma}_h) - \frac{\partial \widetilde{C_t}}{\partial S} = 0 \;, \tag{2.94}$$

which is the volatility that equates the BS delta with the delta of the true, but unknown pricing model, Renault and Touzi (1996). Unfortunately, the hedging volatility is not directly observable.

Renault and Touzi (1996) prove that the bias in this approximation is systematic, when the classical Hull and White (1987) model is the true underlying price process. The bias translates into the following errors in the hedge ratios: for ITM options the use of IV to compute the hedge ratios leads to an underhedged position in the delta, while for OTM options the use of IV leads to an overhedged position. Only for ATM options, in the log-forward moneyness sense, the delta-hedge is perfect. This problem is also demonstrated for both delta and vega risks in a simulation study by Rebonato (1999, Case Study 4.1).

An alternative approach for approximating the unknown delta is to explicitly assume that the smile depends on the underlying asset price S_t. Then one approximates the delta as:

$$\frac{\partial \widetilde{C_t}}{\partial S} = \frac{\partial C^{BS}}{\partial S}(S_t, t, K, T, \widehat{\sigma}) + \frac{\partial C^{BS}}{\partial \widehat{\sigma}}(S_t, t, K, T, \widehat{\sigma}) \frac{\partial \widehat{\sigma}}{\partial S} \;. \tag{2.95}$$

In (2.95) all quantities are known except for $\partial \widehat{\sigma}/\partial S$. It cannot directly be recovered from the IVS, since the IVS is a function in maturity and strikes, *not* in the underlying. One solution would be to use simple conjectures about this quantity, see Derman et al. (1996b) or Derman (1999) and Sect. 3.11 for typical examples. Assuming that volatility is a deterministic function of S_t and t as in Sect. 2.8.1, Coleman et al. (2001) suggest the following approximation to $\partial \widehat{\sigma}/\partial S$. They observe that under this assumption European put and call prices are related to each other through a *reversal* of S and K, and r and δ, respectively, i.e.

$$C(S_t, t, K, T, \widehat{\sigma}, r, \delta) = P(K, t, S_t, T, \widehat{\sigma}, \delta, r) \;, \tag{2.96}$$

where $P(S_t, t, K, T, \sigma, r, \delta)$ denotes the price of a European put option, where – in this order! – current asset price is S_t, t time, strike K, expiry date T, volatility σ and interest rate r, dividend yield δ. In further assuming that a similar relationship holds in terms of IV, they derive

$$\frac{\partial \widehat{\sigma}^C(S_t, t, K, T, r, \delta)}{\partial S} = \frac{\partial \widehat{\sigma}^P(K, t, S_t, T, \delta, r)}{\partial S}, \qquad (2.97)$$

where the superscript C and P denote call and put respectively. Note that the left-hand side in (2.97) is the unknown derivative of IV with respect to the underlying asset, while the right-hand side of (2.97) is a *strike* derivative which can be reconstructed from the IVS. Relation (2.97) is particularly convenient when $r \approx \delta$ (as in the case of futures options), since the switch of both quantities becomes obsolete. In terms of an empirical performance, Coleman et al. (2001) report that this approximation substantially improves the hedges based on a simple constant volatility method.

Another hedging strategy due to Lee (2001) also includes the stochastic volatility case: reconsider the strike-analogue to (2.95). It is given by:

$$\frac{\partial \widetilde{C}_t}{\partial K} = \frac{\partial C^{BS}}{\partial K}(S_t, t, K, T, \widehat{\sigma}) + \frac{\partial C^{BS}}{\partial \widehat{\sigma}}(S_t, t, K, T, \widehat{\sigma})\frac{\partial \widehat{\sigma}}{\partial K}. \qquad (2.98)$$

Multiply (2.95) with S_t and (2.98) with K and sum both equations. Assuming further that $\widetilde{C}_t$ is homogenous of degree one in S_t and K (the BS price C_t^{BS} fulfills this property as can easily be checked), we find:

$$\frac{\partial \widehat{\sigma}}{\partial S} = -\frac{K}{S_t}\frac{\partial \widehat{\sigma}}{\partial K}. \qquad (2.99)$$

Thus the *corrected hedge ratio* is given by:

$$\frac{\partial \widetilde{C}_t}{\partial S} = \frac{\partial C^{BS}}{\partial S}(S_t, t, K, T, \widehat{\sigma}) - \frac{\partial C^{BS}}{\partial \widehat{\sigma}}(S_t, t, K, T, \widehat{\sigma})\frac{K}{S_t}\frac{\partial \widehat{\sigma}}{\partial K}. \qquad (2.100)$$

This delta-hedge has a direct reference to the IVS and can be implemented without estimating an underlying stochastic volatility model. When the smile is negatively skewed, this approach delivers smile dynamics that proxy the so called sticky-moneyness assumption, see the discussion in Sect. 3.11. Moreover, it also gives insights for the results by Renault and Touzi (1996): when the smile is u-shaped, the IV compensated delta overhedges the OTM call, since $\partial \widehat{\sigma}/\partial K$ in the second term of the right-hand side in (2.100) has a positive sign in the OTM regions of calls, Sect. 2.7.

For risk management, other difficulties appear, especially when IV compensated hedge ratios are used. When different BS models apply for different strikes, one may question whether delta and vega risks across different strikes can simply be added to assess the overall risk in the option book: being a certain amount of euro delta long in high strike options, and the same amount delta short in low strike options, need not necessarily imply that the book is eventually delta-neutral. There may be residual delta risk that has to be hedged, even on this aggregate level.

Similarly, the vega risk of the portfolio needs to be carefully assessed. In stress scenarios it is crucial how the IVS is shocked, e.g. whether one shifts the IVS across strikes and time to maturity in an entirely parallel fashion or in more sophisticated ways. This is explored with the dimension reduction techniques developed in Chap. 5, which offer empirical answers to these questions: typically, the most important shocks are due to almost parallel up and down shifts of the IVS. A second source of shocks affects the moneyness slope of the IVS, while a third type influences the moneyness curvature of the IVS or – depending on the modeling approach – its term structure. This will be studied in details in Chap. 5.

2.9.2 Pricing

A next challenge is valuing exotic options. The reason is that even weakly path-dependent options, such as barrier options, require sophisticated volatility specifications. Consider, e.g. an ITM knock-out option with strike K and barrier $L > K$. In this case, explicit valuation formulae are known when the underlying follows a geometric Brownian Motion, Musiela and Rutkowski (1997, Chap. 9). However, which IV should be used for pricing? One could use the IV at the strike K, the one at the barrier L, or some average of both. This problem is the more virulent the more sensitive the exotic option is to volatility.

At this point it becomes clear that, in the presence of the IVS, pricing is not sensible without a self-consistent and reliable model. One way is taken by the stochastic volatility models sketched in Sect. 2.8. Another way, which is much closer to the concept of the IVS, and hence to the topic of this research, is offered by the smile consistent *local* volatility models. These models rely on a volatility function that is directly backed out of prices of plain vanilla options observed in the market. Thus, the exotic option is priced *consistently with the entire IVS*. This is a natural approach, especially when the exotic option is to be hedged with plain vanilla options. It will be the topic in Chap. 3.

2.10 IV as Predictor of Realized Volatility

Forecasting volatility is a major topic in economics and finance: whether in monetary policy making, for investment decisions, in security valuation, or in risk management, a precise assessment of the market's expectations on volatility is inevitable. Consequently, forecasting volatility has received high attention in the past twenty years. One main strand of this literature employs genuine time series models to produce volatility forecasts, and most of these studies rely on the class of autoregressive conditional heteroscedasticity models, which emerged from the initial work of Engle (1982). Another natural methodology is to exploit option IV as predictor for future volatility. Here, we give a short, and by no means comprehensive summary of this second stream of studies. For an excellent survey on this enormous body of literature we refer to Poon and Granger (2003).

In an efficient market, options instantaneously adjust to new information. Thus, IV predictions do not depend on the historical price or volatility series in an adaptive sense, as is typically the case in time series based methods. While this may be seen as a general advantage of IV based methods, there are two methodological caveats: first, the test on the forecasting ability of IV is always a joint test of option market efficiency *and* the option pricing model, which can hardly be disentangled. Second, given the presence of the smile, one either has to restrict the analysis to ATM options or to find an appropriate weighting scheme of IV across different strikes, see Sect. 4.5.2 for a discussion of this point in the context of least squares kernel smoothing of the IVS.

The first to study IV as a predictor for individual stock volatility is Latané and Rendelman (1976), followed by Chiras and Manaster (1978), Schmalensee and Trippi (1978), Beckers (1981), and Lamoureux and Lastrapes (1993). Harvey and Whaley (1992), Christensen and Prabhala (1998), and Shu and Zhang (2003) among others investigate stock market indices. Foreign exchange markets and interest rate futures options are examined by e.g. Jorion (1995), and Amin and Ng (1997), respectively. Most recent research focusses on long memory and fits fractionally integrated autoregressive models for the volatility forecast, Andersen et al. (2003) and Pong et al. (2003). This appears to be a promising line of research.

The overall consensus of the literature regardless of the market and the forecasting horizon under scrutiny appears to be – for an exception see Canina and Figlewski (1993) – that IV based predictors do contain a substantial amount of information on future volatility and are better than (only) time series based methods. At the same time, most authors conclude that IV is a biased predictor. Deeper theoretical insight on the bias is shed by Britten-Jones and Neuberger (2000), see Sect. 3.9, that provide a *model-free* option based volatility forecast. They show that if the IVS does not depend on K and T, but is not necessarily constant, then squared IV is exactly this forecast – however: under the risk neutral measure. Hence, the bias may not be due to model misspecification or measurement errors, but rather due to the way the market prices volatility risk. Thus, research now proposes the volatility risk premium as a possible explanation, Lee (2001) and Bakshi and Kapadia (2003).

2.11 Why Do We Smile?

Ever since the observation of the smile function, research has aimed at explaining this striking deviation from the BS constant volatility assumption. This is achieved by subsequently relaxing the assumptions of the models. Nowadays, the literature comprises a lot of factors possibly responsible for smile and term structure patterns: they range from market microstructure frictions, such as liquidity constraints and transaction costs, to stochastic volatility and Lévy-processes for the underlying asset price process. Although stochastic volatility and asset prices driven by Lévy-processes may be the best understood and most prominent explanations, the empirical literature has been little successful in disentangling the different factors: since IV is a free parameter, it comprises 'expected volatility and *everything else* that affects option supply and demand but is not in the model', Figlewski (1989, p. 13).

It has been conjectured quite early in the literature that stochastic volatility is responsible for the smile effect, but Renault and Touzi (1996) are the first to formally prove this suspicion under the assumption of an underlying Hull and White (1987) model with zero correlation between the two Brownian motions. They show that stochastic volatility necessarily implies a U-shaped smile which attains its minimum for ATM options (in the sense of forward moneyness). A similar conclusion is drawn in stochastic IV models, see Sect. 3.12. The stochastic volatility smile effect is also confirmed in empirical work: for instance, Härdle and Hafner (2000) demonstrate that GARCH-type models considerably reduce the pricing error of options compared with the simple BS model. However, the smile patterns generated by these stochastic volatility models do not appear to match well the ones empirically observed, Heynen (1994). This is also confirmed by Jorion (1988) and Bates (1996) who overall favor jump diffusion models against stochastic volatility. Das and Sundaram (1999) investigate more deeply the implications of the models concerning the shape of the smile and the term structure of the IVS. According to them, stochastic volatility smiles are too shallow, while jump diffusions imply the smile only for short maturity options. Moreover, they prove that jump diffusions always imply an *increasing* term structure of IV. However, empirically, also a decreasing or at least humped term structure is observed, compare Fig. 2.11 for the years 1997 and 2000. Similar results are reported by Tompkins (2001) for a variety of stochastic and jump models. Summing up, it appears that only a combination of jump and stochastic volatility models is sufficiently capable of capturing the stylized facts of the IVS, Bakshi et al. (1997).

As in the literature on the predictive power of IV, new studies seek the reasons for the IVS in long memory in volatility of the underlying process, e.g. Breidt et al. (1998). There is evidence that in particular the upward-sloping term structure of the IVS can strongly be influenced by long memory in volatility, Taylor (2000).

Given that the distributional assumption of normal returns behind the BS model is frequently rejected, see e.g. Ederington and Guan (2002) for an analysis that is based on delta-hedging an option portfolio, processes with a marginal distribution tails heavier than the Gaussian ones are considered. For instance, Barndorff-Nielsen (1997) discusses the inverse Gaussian distribution. This distribution is from the class of generalized hyperbolic distributions, proposed by Eberlein and Keller (1995), Küchler et al. (1999), and Eberlein and Prause (2002) for modeling asset price processes that capture the smile

2.11 Why Do We Smile?

effect. A comprehensive introduction into smile consistent option pricing with Lévy processes is found in Cont and Tankov (2004).

An increasing literature seeks the reasons for the smiling volatility functions in market imperfections. Jarrow and O'Hara (1989) argue that the differences between IV and the historical volatility reflect the transaction costs of the dynamic hedge portfolio. In approximating transaction costs by the bid-ask spread, a similar conclusion is reached by Peña et al. (1999). Within an equilibrium framework, Grossman and Zhou (1996) analyze possible feedback effects from hedging and market illiquidity. In their set-up, portfolio insurance can generate volatility skews. Frey and Patie (2002) show that a market liquidity, which depends on the asset price level, produces smile patterns as are typically observed. This assumption is in line with the experience that large up or down swings in asset prices lead to a decrease of market liquidity.

In Sect. 2.5, it has been argued that also supply and demand conditions may contribute to the shape of the smile. In a recent study, Bollen and Whaley (2003) examine the net buying pressure proxied by the difference of buyer-motivated and seller-motivated contracts. To them, net buying pressure plays an important role for the shape of the IVS in the S&P 500 market. One is tempted to argue that, in an efficient market, a higher increased demand for portfolio insurance should provide incentives to agents to sell options and to replicate them synthetically. However, the presence of short sale and borrowing constraints among investors may make the replication strategy more costly, thereby driving up option prices. Fahlenbrach and Strobl (2002) provide empirical evidence for this argument.

Finally, an interesting explanation for the index smile has recently been put forward: it is a well-known fact that stock smile functions are shallow compared with the smile of index options, Bollen and Whaley (2003). The risk neutral distribution of the index – since it is a (deterministically) weighted average of single stocks – is completely determined by the risk neutral distributions of the single stocks. Branger and Schlag (2004) show that the steepness of the smile is an immediate result of the *dependence structure* of the single stocks in the basket. Moreover, a change in this dependence structure, which has been addressed by Fengler and Schwendner (2004) in the context of pricing multiasset equity options, can have dramatic consequences to the shape of the IVS. Indeed the relation of prices, risk neutral distributions and volatility functions between stock options and basket options is relatively unexplored. First such approaches in this direction are Avellaneda et al. (2002), Bakshi et al. (2003) and Lee et al. (2003).

2.12 Summary

In this chapter we introduced the phenomenon of the IV smile and the IVS: the first part was devoted to an introduction into the BS model for the pricing of contingent claims. Two principles in pricing, the self-financing replication strategy and the probabilistic approach based on the risk neutral measure, were presented. We derived the BS formula for plain vanilla European calls and puts.

The second part treated the concept of IV. We discussed the static properties of the smile function, such as no-arbitrage bounds on the IV slope and the asymptotic behavior of the smile function. A discussion of the general empirical regularities of the IVS observed on equity markets followed. In a first attempt to explain the smile, we presented two typical approaches for relaxing the strict assumptions of the BS model: time-dependent and stochastic volatility. In both frameworks, we arrived at an interpretation of IV as an *average of the squared volatility function*. Then we discussed the challenges in hedging and pricing in the presence of a smile. The chapter concluded with a short summary on the literature that employs IV as a predictor for future stock price fluctuations, and with a complementary section on other possible explanations for the existence of the smile phenomenon.

3

Smile Consistent Volatility Models

3.1 Introduction

The existence of the smile requires the development of new pricing models that capture the static and dynamic distortions of the IVS. One pathway taken first in Merton (1976) and Hull and White (1987) and the subsequent related literature is to add another degree of freedom either to the process of the underlying asset or to the volatility process. This approach has been sketched in Sect. 2.8. The advent of highly liquid option markets, on which large numbers of standardized plain vanilla options are traded at low costs, has reversed the procedure: an emerging strand of literature of so called *smile consistent volatility models* takes the prices of plain vanilla options as given. The aim is to extract information about the asset price dynamics and the volatility directly from the observed option prices and the IVS only, which then is employed to price and hedge other derivative products. The decisive point is that these other derivatives are priced and hedged *relative* to the observed plain vanilla options. The name smile consistent volatility models is derived from the fact that the (European) options priced in these models exactly reproduce the IVS observed empirically.

This approach is justified by at least two empirical facts and one practical consideration: first, option prices and the IVS are readily at hand, if not directly observed. Second, recent studies demonstrate that a large number of option price movements *cannot* be attributed to movements in the underlying or to market microstructure frictions, Bakshi et al. (2000). This leads to the impression that option markets due to their depth and liquidity behave increasingly self-governed by its own supply and demand conditions. This seems to be particularly virulent at the joint expiry dates of futures contracts and options, the 'triple witch days'.

The third and more practical point concerns portfolios of exotic options: necessarily, positions in these options need to be hedged, and in most cases, this hedge will be sought by employing plain vanilla options. A particular strategy is *static hedging*. Unlike dynamic hedging where the hedge is (almost) continuously adjusted, in static hedging, the payoff of the exotic option is replicated by an appropriate portfolio of plain vanilla options. This portfolio remains unaltered up to expiry, Derman et al. (1995), Carr et al. (1998) and Andersen et al. (2002). In this case, pricing exotic derivatives correctly relative to the options that will be used for the hedge is vital for its accuracy.

In achieving these goals, two main lines of models have emerged: first, *local volatility models* and their most recent stochastic ramifications, and second, *stochastic implied volatility models*. In both approaches, the parameters are obtained from a calibration of the model to a cross-section of option prices. Furthermore, both models allow for preference-free derivative valuation (except for the stochastic local volatility models), since within each modeling framework the market is complete. Thus they do not require additional assumptions on the market price of risk.

The concept central to local volatility models is the *local volatility surface* (LVS). Unlike the IVS, which is a *global* measure of volatility, as can be understood from the averaging concept usually attributed to it, the LVS is a *local* measure in the sense that it gives a volatility forecast for a pair of a particular strike and a particular expiry date (K, T). In this framework instantaneous volatility is not necessarily a deterministic function of time and asset prices, it may perfectly be stochastic. However, in the derivation of the LVS all sources of risk in the stochastic volatility are integrated out, which leave as only risky element the fluctuations of the asset price, Derman and Kani (1998). By its local nature, the LVS – opposite to the IVS – is the correct input parameter for pricing models. Most recently, a number of studies try to circumvent the static implications of the LVS in moving towards stochastic local volatility models.

Stochastic IV models explicitly allow for a stochastic setting. However, the additional state variable is not introduced in the instantaneous volatility function as in the classical stochastic volatility literature, but tied to a stochastic IV. Ultimately, of course, this also implies a stochastic instantaneous volatility. Since plain vanilla options are still priced via the BS formula using the contemporaneous realization of the IV, volatility risk is tradable and the market complete. This allows for a preference-free valuation of contingent claims.

In this chapter, we aim at giving a comprehensive review on the current state of literature of local volatility and stochastic IV models (see also Skiadopoulos (2001) for an excellent review). First, the notion of local volatility as pioneered by Dupire (1994) and Derman and Kani (1998) is presented. In Sect. 3.3 we relate local volatility to observed option prices. The central result will be the so called *Dupire* formula. An alternative path to the Dupire formula is given in Sect. 3.4. Section 3.5 establishes the link between local volatility and IV. The theoretical part of local volatility is deepened in Sect. 3.8, which develops the local volatility as an expected value of instantaneous volatility under the K-*strike* and T-*maturity forward risk-adjusted measure*. Section 3.9 shows how model-free (implied) volatility forecasts can be extracted from option data, Britten-Jones and Neuberger (2000). Finally, a variety of specific models for pricing and extracting local volatility are presented: the main focus is on implied trees, but we also inspect methods motivated from a continuous time setting. Stochastic local volatility models are also considered. The chapter concludes by presenting the stochastic IV model and its properties in Sect. 3.12.

3.2 The Theory of Local Volatility

The concept of *local* volatility (also called forward volatility) was introduced by Dupire (1994), and further developed in Derman and Kani (1998). Intuitively one may think about local volatility, denoted by $\sigma_{K,T}$, as the market's consensus of *instantaneous* volatility for a market level K at some future date T. The ensemble of such estimates for a collection of market levels and future dates is called the *local volatility surface* (LVS). Since it is implied from observed option prices, the LVS gives the fair value of the asset price volatility for future market levels and times. Note the difference to the concept of IV which under certain conditions is thought of as the market's estimate of expected *average* volatility through the life time of the option, Sect. 2.8.

To make the concept of local volatility more precise, we reconsider the continuous-time economy with a trading interval $[0, T^*]$, where $T^* > 0$. Let $(\Omega, \mathcal{F}, P)$ be a probability space, on which at least one Brownian motion $\left(W_t^{(0)}\right)_{0 \leq t \leq T^*}$, but possibly also more Brownian motions, are defined. As usually, P is the objective probability measure and information is revealed by a filtration $(\mathcal{F}_t)_{0 \leq t \leq T^*}$. The asset price $(S_t)_{0 \leq t \leq T^*}$ is modelled by a $(\mathcal{F}_t)_{0 \leq t \leq T^*}$-adapted stochastic process, driven by the SDE

$$\frac{dS_t}{S_t} = \mu(S_t, t) \, dt + \sigma(S_t, t, \cdot) \, dW_t^{(0)}, \qquad (3.1)$$

where $\mu(\cdot, \cdot)$ denotes the instantaneous drift. We assume that the instantaneous volatility $\left(\sigma(S_t, t, \cdot)\right)_{0 \leq t \leq T^*}$ follows some $(\mathcal{F}_t)_{0 \leq t \leq T^*}$-adapted stochastic process possibly depending on S_t, the history of S_t or on other state variables. This arbitrary dependence is meant with the '··-notation'. Finally, we assume absence of arbitrage, which implies the existence of some risk neutral measure $Q \in \mathcal{Q}$ equivalent to P, under which the discounted asset price $(\widetilde{S}_t)_{0 \leq t \leq T^*}$ is a martingale. If the martingale measure is not unique, we think about Q as the risk-neutral measure 'the market has agreed upon', i.e. some *market measure*, see Cont (1999) or Björk (1998, p. 150) for a discussion of this notion. It is also assumed that the entire spectrum of European plain vanilla call prices $C_t(K, T)$, which are priced under Q, are given for any strike K and maturity date T: $\mathcal{G} = \{C_t(K, T), \ K \geq 0, \ 0 \leq T \leq T^*\}$.

The local variance $\sigma_{K,T}^2(S_t, t)$ is defined as the risk-neutral expectation of squared instantaneous volatility conditional on $S_T = K$, and time t information $\mathcal{F}_t$:

$$\sigma_{K,T}^2(S_t,t) \stackrel{\text{def}}{=} \mathsf{E}^Q\{\sigma^2(S_T,T,\cdot)|S_T = K, \mathcal{F}_t\}, \tag{3.2}$$

where $\mathsf{E}^Q(\cdot)$ is the expectation operator under the measure Q. Then local volatility is given by:

$$\sigma_{K,T} \stackrel{\text{def}}{=} \sqrt{\sigma_{K,T}^2}. \tag{3.3}$$

This definition of local volatility has two implications: first, the use of the market's view on future volatility expressed by the expectation operator clarifies that all sources of risk from the stochastic volatility are integrated out. Instead, the evolution of volatility is compressed into a single function that is deterministic in S_t and t. To put it differently, the concept of local volatility presumes – as time elapses – that *the instantaneous volatility will evolve entirely along today's market expectations sublimated in the local volatility function*. Therefore, within a local volatility framework, for some market level $K = S_t$ at $T = t$, the instantaneous volatility is:

$$\sigma(S_t,t) = \sigma_{S_t,t}(S_t,t), \tag{3.4}$$

and the asset price is driven by:

$$\frac{dS_t}{S_t} = \mu(S_t,t)\,dt + \sigma_{S_t,t}(S_t,t)\,dW_t^{(0)}. \tag{3.5}$$

It is precisely this feature which allows to use the LVS directly in the generalized BS PDE (2.67) as a (market implied) volatility function to price exotic or illiquid options, since the market remains complete and derivative valuation preference-free. In this sense local volatility ensures that these other options are correctly priced *relative* to the observed plain vanilla options. This simplicity, however, comes at a cost: whereas (3.1) includes *all* stochastic volatility models, (3.5) is a one-factor diffusion with a deterministic (though possibly very complicated) volatility function. It can be questioned whether this delivers an adequate description of asset price behavior, Hagan et al. (2002) and Ayache et al. (2004), and Sect. 3.11.

At this point it is useful to revoke the similarity of local volatility and the forward rate. The key insight by Derman and Kani (1998) is that local volatilities are constructed in an analogous way as forward rates in the theory of interest rates. They play the same role: in the same way as bond prices that are computed from the forward rate curve match their market prices, so do option prices calculated from the local volatility function. Just as the first is extracted from bond prices and then employed to correctly price derivatives or other bonds, so is the latter. Using forward rates does not imply the believe that they are the right predictors for interest rates. The same applies to local volatility with respect to future instantaneous volatility. However, despite this fact, forward rates are the relevant quantities for a bond trader in this context.

Second, as a minor and obvious implication, if instantaneous volatility is deterministic in spot and time, i.e. $\sigma(S_t,t,\cdot) \stackrel{\text{def}}{=} \sigma(S_t,t)$, both concepts, instantaneous *and* local variance, coincide:

$$\sigma_{K,T}^2(S_t,t) \stackrel{\text{def}}{=} \mathsf{E}^Q\{\sigma^2(S_T,T,\cdot)|S_T=K,\mathcal{F}_t\}$$
$$= \mathsf{E}^Q\{\sigma^2(S_T,T)|S_T=K,\mathcal{F}_t\} = \sigma^2(K,T) \,. \qquad (3.6)$$

In this case, instantaneous volatility evolves along the static local volatility function, since the right-hand side is independent of S and t.

Derman and Kani (1998) further characterize the local variance in showing that it can be represented as the risk-adjusted expectation of the future instantaneous variance at time T:

$$\sigma_{K,T}^2(S_t,t) = \mathsf{E}^{(K,T)}\{\sigma^2(S_T,T,\cdot)|\mathcal{F}_t\} \,, \qquad (3.7)$$

where the expectation is now taken with respect to a new measure, which is called the *K-strike* and *T-maturity forward risk-adjusted measure*. This is again in analogy with the theory of forward rates: here, the forward rate is obtained by taking the expectation of the short rate under the T-maturity forward measure, Jamshidian (1993). The derivation of (3.7) will be delayed until Sect. 3.8.

Clearly, for pricing, the assumption that the only source of risk is the asset price may be considered as a drawback. It may be good for markets in which asset prices and volatility are strongly correlated, as is commonly seen in equity markets, but can be questioned for foreign exchange markets. The dynamic hedging performance of the deterministic local volatility models is criticized, Hagan et al. (2002). Furthermore, they do not provide a genuine explanation for the smile phenomenon, but rather overstretch the ordinary BS world, Ayache et al. (2004). This, however, does not appear to diminish their significance in pricing exotic derivatives in practice. In order to meet this criticism and to improve the hedging performance, recent work aims at relaxing the deterministic framework and moves towards a stochastic theory of local volatility.

3.3 Backing the LVS Out of Observed Option Prices

As forward rates are intricately linked to observed bond prices, so is local volatility to observed option prices. Here, we show how the local volatility function is recovered from the set of European call option prices $\mathcal{G}$. The exposition follows Derman and Kani (1998).

Under the equivalent martingale measure asset prices follow the SDE:

$$\frac{dS_t}{S_t} = (r-\delta)\,dt + \sigma(S_t,t,\cdot)\,d\overline{W}_t^{(0)} \,, \qquad (3.8)$$

where $\overline{W}_t^{(0)}$ denotes the Brownian motion, which drives the asset price, under the risk neutral measure Q. The interest rate and the continuously compounded dividend yield are denoted by r and δ, respectively.

3 Smile Consistent Volatility Models

By the martingale property, the calls are priced by

$$C_t(K,T) = e^{-r\tau} \mathsf{E}^Q\{(S_T - K)^+ | \mathcal{F}_t\}, \qquad (3.9)$$

where $\tau \stackrel{\text{def}}{=} T - t$. Taking the left side first order derivative with respect to K yields

$$D^- C_t(K,T) = -e^{-r\tau} \mathsf{E}^Q\{\mathbf{1}(S_T > K) | \mathcal{F}_t\}. \qquad (3.10)$$

Differentiating again we recover

$$\frac{\partial^2 C_t(K,T)}{\partial K^2} = e^{-r\tau} \mathsf{E}^Q\{\delta_K(S_T) | \mathcal{F}_t\}, \qquad (3.11)$$

where $\delta_{x_0}(\cdot)$ denotes the Dirac delta function, which is defined by the property $\int f(x)\,\delta_{x_0}(x)\,dx = f(x_0)$ for a smooth function f. The derivative $\frac{\partial^2}{\partial K^2}(S_T - K)^+ = \delta_K(S_T)$ is defined in a distributional sense, see Hormander (1990) for a proper mathematical formulation of distributions. Note that Equation (3.11) shows yet another derivation of the state price density, Sect. 2.4.

In a second step we take the derivatives of (3.9) with respect to T:

$$\frac{\partial C_t(K,T)}{\partial T} = -rC_t(K,T) + e^{-r\tau}\frac{\partial}{\partial T}\mathsf{E}^Q\{(S_T - K)^+ | \mathcal{F}_t\}. \qquad (3.12)$$

To evaluate the right-hand side of (3.12) we apply a generalization of the Itô formula to the convex function $(S_T - K)^+$, called *Tanaka-Meyer formula*, Appendix (B.14). This yields

$$d(S_T - K)^+ = \mathbf{1}(S_T > K)\,dS_T + \frac{1}{2}S_T^2\,\sigma^2(S_T, T, \cdot)\,\delta_K(S_T)\,dT. \qquad (3.13)$$

Taking expectations in (3.13) together with the asset price dynamics (3.8) yields:

$$d\mathsf{E}^Q\{(S_T - K)^+ | \mathcal{F}_t\} = \\ (r-\delta)\mathsf{E}^Q\{S_T\mathbf{1}(S_T > K)|\mathcal{F}_t\}\,dT + \mathsf{E}^Q\left\{\tfrac{1}{2}S_T^2\sigma^2(S_T,T,\cdot)\delta_K(S_T)|\mathcal{F}_t\right\}dT. \qquad (3.14)$$

The first term in the previous equation can be split into

$$\mathsf{E}^Q\{S_T\mathbf{1}(S_T > K)\} = \mathsf{E}^Q\{(S_T - K)^+\} + K\mathsf{E}^Q\{\mathbf{1}(S_T > K)\}. \qquad (3.15)$$

Plugging this into (3.14) and using (3.9) and (3.10), one obtains:

$$\frac{\partial}{\partial T}\mathsf{E}^Q\{(S_T - K)^+ | \mathcal{F}_t\} = e^{r\tau}(r - \delta)\left\{C_t(K,T) - K\frac{\partial C_t(K,T)}{\partial K}\right\} \\ + \frac{1}{2}K^2\mathsf{E}^Q\{\sigma^2(S_T, T, \cdot)\delta_K(S_T) | \mathcal{F}_t\}. \qquad (3.16)$$

3.3 Backing the LVS Out of Observed Option Prices

By the law of iterated expectations the last term in (3.16) can be rewritten as:

$$\mathsf{E}^Q\{\sigma^2(S_T,T,\cdot)\delta_K(S_T)|\mathcal{F}_t\} = \mathsf{E}^Q[\mathsf{E}^Q\{\sigma^2(S_T,T,\cdot)\delta_K(S_T)|S_T=K,\mathcal{F}_t\}|\mathcal{F}_t]$$
$$= \mathsf{E}^Q\{\sigma^2(S_T,T,\cdot)|S_T=K,\mathcal{F}_t\}\mathsf{E}^Q\{\delta_K(S_T)|\mathcal{F}_t\}\,.$$
(3.17)

Thus, inserting (3.16) and (3.17) into (3.12), we find that

$$\frac{\partial C_t(K,T)}{\partial T} = -\delta C_t(K,T) - (r-\delta)K\frac{\partial C_t(K,T)}{\partial K}$$
$$+ \frac{1}{2}K^2\frac{\partial^2 C_t(K,T)}{\partial K^2}\mathsf{E}^Q\{\sigma^2(S_T,T,\cdot)|S_T=K,\mathcal{F}_t\}\,. \quad (3.18)$$

Solving for the volatility function $\mathsf{E}^Q\{\sigma^2(T,\cdot)|S_T=K,\mathcal{F}_t\}$ yields:

$$\sigma^2_{K,T}(S_t,t) = 2\,\frac{\frac{\partial C_t(K,T)}{\partial T} + \delta C_t(K,T) + (r-\delta)K\frac{\partial C_t(K,T)}{\partial K}}{K^2\frac{\partial^2 C_t(K,T)}{\partial K^2}}\,, \quad (3.19)$$

where $\sigma^2_{K,T}(S_t,t) \stackrel{\text{def}}{=} \mathsf{E}^Q\{\sigma^2(S_T,T,\cdot)|S_T=K,\mathcal{F}_t\}$. This is the *Dupire formula*, Dupire (1994). The Dupire formula gives a representation of the local volatility function completely in terms of observed call prices and their derivatives.

It remains to show that local volatility $\sigma_{K,T}(S_t,t) \stackrel{\text{def}}{=} \sqrt{\sigma^2_{K,T}(S_t,t)}$ is indeed a real number. This can be seen by the following observations: the denominator of (3.19) is positive by no-arbitrage, since the transition probability must be positive on the entire support. Positiveness of the numerator is obtained by a portfolio dominance arguments similar to those in Merton (1973), see Andersen and Brotherton-Ratcliffe (1997). We have:

$$e^{\delta\varepsilon}C_t(Ke^{(r-\delta)\varepsilon},T+\varepsilon) \geq C_t(K,T) \quad (3.20)$$

for $\varepsilon > 0$. A Taylor series expansion of order one in the neighborhood of $\varepsilon = 0$ yields:

$$\frac{\partial C_t(K,T)}{\partial T} + \delta C_t(K,T) + (r-\delta)K\frac{\partial C_t(K,T)}{\partial K} \geq 0\,. \quad (3.21)$$

Thus it is verified that local volatility $\sqrt{\sigma^2_{K,T}(S_t,t)}$ is indeed a real number.

The result in (3.19) holds irrespective of the assumptions made on the process $(\sigma(S_t,t,\cdot))_{0\leq t\leq T^*}$. If instantaneous volatility is assumed to be deterministic, as Dupire (1994) did in his original work, the expectation operator can be dropped. In this case, it can be shown that the diffusion is completely characterized by the (risk neutral) transition probability, Sect. 3.4.

It is interesting to interpret (3.19) in terms of trading strategies: the numerator is related to an infinitesimal calendar spread, while the denominator contains the position of an infinitesimal butterfly spread. Thus, from a trading perspective local volatility is linked to the ratio of both types of spreads. These considerations imply that local volatility can be locked by appropriate trading strategies as one can lock the forward rates in trading bonds. This is discussed in Derman et al. (1997).

3.4 The *dual* PDE Approach to Local Volatility

There is a remarkable second approach for deriving the Dupire formula (3.19), Dupire (1994). This approach directly builds on the transition probability. While in general it is not possible to recover the dynamics of the asset price process from the transition probability, there is one exception: if one considers one-factor diffusions only, i.e. if one initially assumes instantaneous volatility to be a deterministic function in the asset price and time. The reason is that there exists a *dual* or *adjoint* PDE to the BS PDE (2.13) which has, instead of S and t, K and T as independent variables.

Assume now that under the risk neutral measure Q the asset price dynamics are given by:

$$\frac{dS_t}{S_t} = (r - \delta)\,dt + \sigma(S_t, t)\,d\overline{W}_t^{(0)}, \qquad (3.22)$$

where the notation stays as before except that $\sigma(S_t, t)$ is deterministic.

It is well known that the risk neutral transition probability $\phi(K, T | S_t, t) \stackrel{\text{def}}{=} e^{r\tau} \partial^2 C_t(K, T)/\partial K^2$, introduced in (2.38), satisfies the BS PDE (2.13) with terminal condition:

$$\phi(K, T | S_T, T) = \delta_K(S_T). \qquad (3.23)$$

However, it also satisfies the *Fokker-Planck* or *forward Kolmogorov* PDE, see appendix (B.24). This yields:

$$\frac{\partial \phi(K', T | S_t, t)}{\partial T} = \frac{1}{2} \frac{\partial^2}{\partial (K')^2} \left\{ \sigma^2(K', T)(K')^2 \phi(K', T | S_t, t) \right\}$$
$$- \frac{\partial}{\partial K'} \left\{ (r - \delta) K' \phi(K', T | S_t, t) \right\} \qquad (3.24)$$

for *fixed* S_t and t, over all maturities T and strikes K' with initial condition:

$$\phi(K', t | S_t, t) = \delta_S(K'). \qquad (3.25)$$

To derive the Dupire formula one substitutes for $\phi(K, T | S_t, t)$. Evaluating the first term in (3.24) yields:

$$\frac{\partial \phi(K', T | S_t, t)}{\partial T} = \frac{\partial}{\partial T} \left\{ e^{r\tau} \frac{\partial^2 C_t(K', T)}{\partial (K')^2} \right\}$$
$$= re^{r\tau} \frac{\partial^2 C_t(K', T)}{\partial (K')^2} + e^{r\tau} \frac{\partial^2}{\partial (K')^2} \frac{\partial C_t(K', T)}{\partial T}. \qquad (3.26)$$

Next, we find the term on the right-hand side in (3.24):

$$\frac{\partial}{\partial K'}\{(r-\delta)K'\phi(K',T|S_t,t)\} = (r-\delta)e^{r\tau}\frac{\partial}{\partial K'}\left(K'\frac{\partial^2 C_t}{\partial (K')^2}\right). \quad (3.27)$$

Thus, (3.24) results in

$$r\frac{\partial^2 C_t(K',T)}{\partial (K')^2} + \frac{\partial^2}{\partial (K')^2}\frac{\partial C_t(K',T)}{\partial T} = \frac{1}{2}\frac{\partial^2}{\partial^2 K'}\left\{\sigma^2(K',T)(K')^2\frac{\partial^2 C_t}{\partial (K')^2}\right\}$$
$$- (r-\delta)\frac{\partial}{\partial K'}\left(K'\frac{\partial^2 C_t}{\partial (K')^2}\right). \quad (3.28)$$

Integrating (3.28) twice from K to infinity, yields:

$$rC_t(K,T) + \frac{\partial C_t(K,T)}{\partial T} - \frac{1}{2}K^2\sigma^2(K,T)\frac{\partial^2 C_t(K,T)}{\partial K^2}$$
$$+ (r-\delta)K\frac{\partial C_t(K,T)}{\partial K} - (r-\delta)C_t(K,T) = 0, \quad (3.29)$$

under the following assumptions: given that the payoff function of a call is $\psi = (S-K)^+$, the call price and its first and second order derivatives as functions of the strike must tend to zero as K tends to infinity. More precisely, we require that

$$C_t(K,T), \; K\frac{\partial C_t}{\partial K}, \; K^2\frac{\partial^2 C_t}{\partial K^2}, \; K^2\frac{\partial^3 C_t}{\partial K^3} \to 0 \text{ as } K \to \infty. \quad (3.30)$$

Note that (3.30) has implications for the tail behavior of the (risk-neutral) transition density $\phi(K,T|S_t,t)$, which must be $o(K^{-2})$. With regard to the BS pricing function, it is evident that the assumptions (3.30) hold given the exponential decay of the (log-normal) transition density, see Equation (2.40).

From (3.29) the Dupire formula (3.19) is readily received by solving for $\sigma^2(K,T)$. The final arguments are the same as given in Sect. 3.3 following Equation (3.19). Uniqueness is proved in Derman and Kani (1994b).

3.5 From the IVS to the LVS

An open question up to now is how the IVS and the LVS can be linked. This would be desirable from two points of view: first, in a static situation, one could immediately recover the LVS, which in principle is unobservable, from the easily observable IVS. Second, in a dynamic context, it adds additional value to the dynamical description of the IVS, for instance in terms of the semiparametric factor model, Chap. 5: given a low-dimensional description of the *IVS dynamics*, a representation of the Dupire formula in terms of IV could be exploited to yield the corresponding *LVS dynamics*. This may help improve the hedging performance of local volatility models. Another obvious application could be stress tests for portfolios of exotic options. Here, one could simulate the IVS within the semiparametric factor model. IVS scenarios are then converted into LVS scenarios. The latter are the basis for correctly pricing the exotic options in the portfolio and computing a value at risk measure.

The central idea to obtain such an *IV counterpart* of the Dupire formula is to exploit the BS formula as an analytical vehicle, Andersen and Brotherton-Ratcliffe (1997) and Dempster and Richards (2000). More precisely, we insert the BS formula and its derivatives into the Dupire formula (3.19). In doing so, the BS formula is interpreted as if IV depended on K and T as one empirically observes on the markets, i.e. we assume:

$$C^{BS}(S_t, t, K, T, \sigma, r, \delta) = C^{BS}(S_t, t, K, T, \widehat{\sigma}(K,T), r, \delta) \,. \qquad (3.31)$$

Furthermore, we maintain our assumption that local volatility is a deterministic function.

Applying the chain rule of differentiation, we obtain for the numerator of the Dupire formula, suppressing the dependence of $\widehat{\sigma}$ on K and T:

$$2\left\{\frac{\partial C_t^{BS}}{\partial T} + \frac{\partial C_t^{BS}}{\partial \widehat{\sigma}}\frac{\partial \widehat{\sigma}}{\partial T} + \delta C_t^{BS} + (r-\delta)K\left(\frac{\partial C_t^{BS}}{\partial K} + \frac{\partial C_t^{BS}}{\partial \widehat{\sigma}}\frac{\partial \widehat{\sigma}}{\partial K}\right)\right\}. \qquad (3.32)$$

Now, the analytical expressions for the BS formula (2.23) and its K- and T-derivatives in (2.33) and in (2.36) are inserted. Most of the terms cancel out. The strategy in the further derivation is to express the remaining terms using the volatility derivative, the vega (2.30). This yields:

$$2\frac{\partial C_t^{BS}}{\partial \widehat{\sigma}}\left\{\frac{\widehat{\sigma}}{2\tau} + \frac{\partial \widehat{\sigma}}{\partial T} + (r-\delta)K\frac{\partial \widehat{\sigma}}{\partial K}\right\}. \qquad (3.33)$$

In differentiating the denominator, we get:

$$K^2\left\{\frac{\partial^2 C_t^{BS}}{\partial K^2} + 2\frac{\partial^2 C_t^{BS}}{\partial K \partial \widehat{\sigma}}\frac{\partial \widehat{\sigma}}{\partial K} + \frac{\partial^2 C_t^{BS}}{\partial \widehat{\sigma}^2}\left(\frac{\partial \widehat{\sigma}}{\partial K}\right)^2 + \frac{\partial C_t^{BS}}{\partial \widehat{\sigma}}\frac{\partial^2 \widehat{\sigma}}{\partial K^2}\right\}. \qquad (3.34)$$

Once again one substitutes the analytical BS derivatives and introduces into each term the BS vega. This results in

$$K^2\frac{\partial C_t^{BS}}{\partial \widehat{\sigma}}\left\{\frac{1}{K^2\widehat{\sigma}\tau} + \frac{2d_1}{K\widehat{\sigma}\sqrt{\tau}}\frac{\partial \widehat{\sigma}}{\partial K} + \frac{d_1 d_2}{\widehat{\sigma}}\left(\frac{\partial \widehat{\sigma}}{\partial K}\right)^2 + \frac{\partial^2 \widehat{\sigma}}{\partial K^2}\right\}. \qquad (3.35)$$

Finally, collecting the numerator (3.33) and the denominator (3.35) shows:

$$\sigma^2_{K,T}(S_t, t) = \frac{\frac{\widehat{\sigma}}{\tau} + 2\frac{\partial \widehat{\sigma}}{\partial T} + 2K(r-\delta)\frac{\partial \widehat{\sigma}}{\partial K}}{K^2\left\{\frac{1}{K^2\widehat{\sigma}\tau} + 2\frac{d_1}{K\widehat{\sigma}\sqrt{\tau}}\frac{\partial \widehat{\sigma}}{\partial K} + \frac{d_1 d_2}{\widehat{\sigma}}\left(\frac{\partial \widehat{\sigma}}{\partial K}\right)^2 + \frac{\partial^2 \widehat{\sigma}}{\partial K^2}\right\}}. \qquad (3.36)$$

This is the Dupire formula in terms of the IVS and its derivatives.

Obviously, this approach does not provide a theory unifying both concepts. This requires more careful treatment, and – up to now – has only been achieved in certain asymptotic situations, Berestycki et al. (2002) and Sect. 3.6. Rather, it is an ad hoc, but successful procedure to link the unobservable LVS with the IVS. Given (3.36) and (3.39), one estimates the IVS and plugs it into (3.36), which yields an estimate of the LVS. The LVS is then used as input factor in pricing algorithms, e.g. in finite difference schemes that solve the generalized BS PDE, Andersen and Brotherton-Ratcliffe (1997) and Randall and Tavella (2000).

For a deeper understanding, of formula (3.36) it is instructive to inspect the situation of no strike-dependence in the IVS. In this case all derivatives with respect to K vanish and (3.36) reduces to

$$\sigma_T^2(t) = \widehat{\sigma} + 2\tau\widehat{\sigma}\frac{\partial \widehat{\sigma}}{\partial T}, \tag{3.37}$$

which implies:

$$\widehat{\sigma}^2 = \frac{1}{\tau}\int_t^T \sigma_T^2(u)\,du. \tag{3.38}$$

Thus, this situation specializes to our previous interpretation of squared IV as *average* squared (local) volatility through the life time of an option, Sect. 2.8. It demonstrates that IV is a *global* measure of volatility, while the LVS is a *local* measure of volatility giving a volatility forecast for a particular pair (K,T).

For a graphical illustration of the LVS, we derive another version of (3.36) in terms of the forward moneyness measure $\kappa_f \stackrel{\text{def}}{=} K/F_t = K/\{e^{(r-\delta)\tau}S_t\}$ and time to maturity τ. After some manipulations, the LVS is given by

$$\sigma_{\kappa_f,\tau}^2(S_t,t) = \frac{\widehat{\sigma}^2 + 2\widehat{\sigma}\tau\frac{\partial \widehat{\sigma}}{\partial \tau}}{1 + 2\kappa_f\sqrt{\tau}d_1\frac{\partial \widehat{\sigma}}{\partial \kappa_f} + d_1 d_2 (\kappa_f)^2\tau\left(\frac{\partial \widehat{\sigma}}{\partial \kappa_f}\right)^2 + \widehat{\sigma}\tau(\kappa_f)^2\frac{\partial^2 \widehat{\sigma}}{\partial \kappa_f^2}}, \tag{3.39}$$

where d_1 and d_2 are interpreted as $d_1 = -\ln(\kappa_f)/(\widehat{\sigma}\sqrt{\tau}) + 0.5\,\widehat{\sigma}\sqrt{\tau}$ and $d_2 = d_1 - \widehat{\sigma}\sqrt{\tau}$.

In Fig. 3.1, we present an estimate of the LVS based on the moneyness representation of the Dupire formula. The derivatives of the IVS are estimated as derivatives of local polynomials of order two which are used to smooth the IVS, see Sect. 4.3 for a description of this procedure. Due to the different scales, the LVS appears to be flatter than the IVS at first glance. As we show in Fig. 3.2, which displays slices from both functions at the maturity of one and three months, this impression is erroneous: it is the LVS which is steeper than the IVS (leaving out the spiky short term local volatilities). Derman et al. (1996b) report as an empirical regularity in equity markets that the smile of the local volatility is approximately two times steeper than the IV smile. They

58   3 Smile Consistent Volatility Models

IV ticks and IVS: 20000502

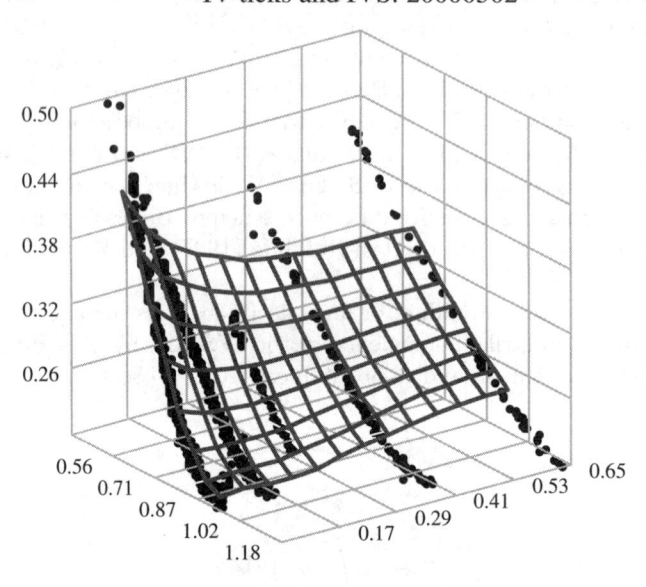

LVS: 20000502

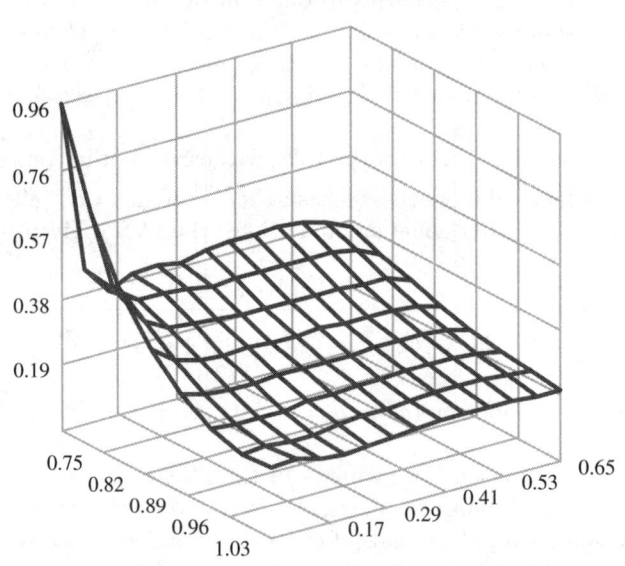

Fig. 3.1. *Top panel*: DAX option IVS on 20000502. IV observations are displayed as *black* dots; the surface estimate is obtained from a local quadratic estimator with localized bandwidths. *Bottom panel*: LVS on 20000502; obtained from the IVS given in the top panel via the moneyness representation of the Dupire formula (3.39)
 SCMlvs.xpl

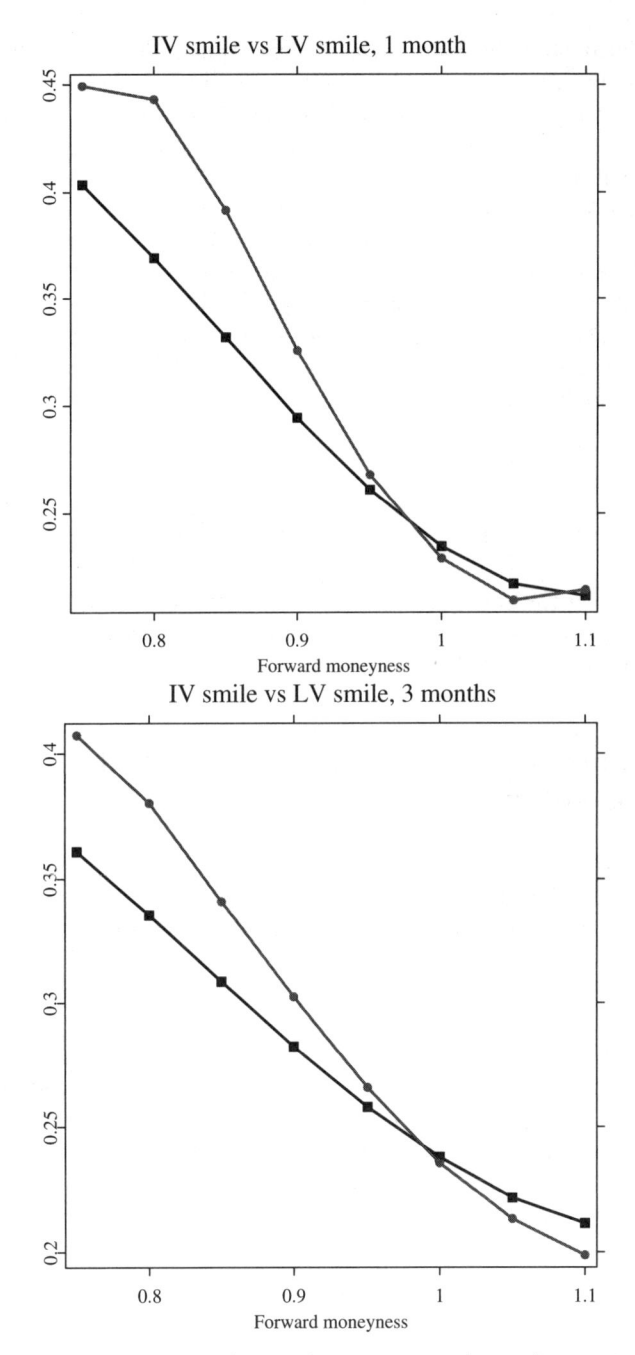

Fig. 3.2. DAX option implied (*squares*) versus local (*circles*) volatility smiles for one month and three months to expiry respectively on 20000502 taken as slices from Fig. 3.1 SCMlvs.xpl

call this relationship the *two-times-IV-slope-rule* for local volatility. Using a recent result by Berestycki et al. (2002) we shall prove in Sect. 3.7 that this conjecture can be made more precise for short maturity ATM options.

In fact, there are a large number of other procedures to reconstruct the LVS. They will be separately surveyed in Sect. 3.10, among them the *implied tree approaches*. Another important stream of literature calls for a more formal mathematical treatment and recovers the LVS from the Dupire formula or the dual PDE in terms of an *(ill-posed) inverse problem*.

As a final cursory remark, note that Equation (3.35), if we ignore the initial K^2-term, is nothing but an expansion of the state price density in terms of the BS vega, the smile and its first and second order derivatives, see the discussion in Sect. 2.4, pp. 18:

$$\phi(K,T|S_t,t) = e^{-\delta\tau} S_t \sqrt{\tau} \varphi(d_1) \tag{3.40}$$

$$\times \left\{ \frac{1}{K^2 \widehat{\sigma} \tau} + \frac{2d_1}{K\widehat{\sigma}\sqrt{\tau}} \frac{\partial \widehat{\sigma}}{\partial K} + \frac{d_1 d_2}{\widehat{\sigma}} \left(\frac{\partial \widehat{\sigma}}{\partial K} \right)^2 + \frac{\partial^2 \widehat{\sigma}}{\partial K^2} \right\},$$

In estimating the smile and its derivatives, expression in (3.40) may serve as a vehicle to recover the state price density, see Huynh et al. (2002) and Brunner and Hafner (2003) for details.

3.6 Asymptotic Relations Between Implied and Local Volatility

Recent research has identified situations in which the relation between implied and local volatility can be established more exactly. These results are of asymptotic nature and more general than those stated so far, since they allow the local volatility to be strike-dependent. More precisely, Berestycki et al. (2002) show that near expiry, IV can be represented as the *spatial harmonic mean* of local volatility. The key consequence of this result is that the IVS can be extended up to $\tau = 0$ as a *continuous function*. This can be exploited in the calibration of local volatility models, Sect. 3.10.3. Additionally, they prove that the representation (3.38), i.e. squared IV as an average of squared local volatility, holds also for deep OTM options under certain assumptions.

To obtain their results, Berestycki et al. (2002) assume that local volatility is deterministic. As noted in (3.6), this implies

$$\sigma^2_{K,T}(S_t,t) = \sigma^2(K,T), \tag{3.41}$$

i.e. local volatility *is* the instantaneous volatility function for all $S_t = K$ and $t = T$. Further they transform the Dupire formula, into the (inverse) log-forward moneyness space, similarly as we have done to derive the forward moneyness representation for the empirical demonstration in the previous section. Define

3.6 Asymptotic Relations Between Implied and Local Volatility

$$x \stackrel{\text{def}}{=} -\ln \kappa_f = \ln(S_t/K) + (r-\delta)\tau. \qquad (3.42)$$

Straightforward calculations show that this transforms the IV counterpart of Dupire (3.36) into the following quasilinear parabolic PDE of IV, where we suppress the dependence of IV on x and τ:

$$2\tau\widehat{\sigma}\frac{\partial\widehat{\sigma}}{\partial\tau} + \widehat{\sigma}^2 - \sigma^2(x,\tau)\left(1 - \frac{x}{\widehat{\sigma}}\frac{\partial\widehat{\sigma}}{\partial x}\right)^2$$
$$- \sigma^2(x,\tau)\tau\widehat{\sigma}\frac{\partial^2\widehat{\sigma}}{\partial x^2} + \frac{1}{4}\sigma^2(x,\tau)\tau^2\widehat{\sigma}^2\left(\frac{\partial\widehat{\sigma}}{\partial x}\right)^2 = 0. \qquad (3.43)$$

To gain an insight into the nature of this first result, consider the following: let $\widehat{\sigma}(x,0)$ be the unique solution to the PDE at $\tau = 0$. Then (3.43) reduces to

$$\widehat{\sigma}^2(x,0) - \sigma^2(x,0)\left\{1 - \frac{x}{\widehat{\sigma}(x,0)}\frac{\partial\widehat{\sigma}(x,0)}{\partial x}\right\}^2 = 0. \qquad (3.44)$$

By simple calculations it is seen that the solution is:

$$\widehat{\sigma}(x,0) = \left\{\int_0^1 \frac{ds}{\sigma(sx,0)}\right\}^{-1} = \left\{\frac{1}{x}\int_0^x \frac{dy}{\sigma(y,0)}\right\}^{-1}, \qquad (3.45)$$

where the second more familiar representation is obtained by the variable substitution $y = sx$ for $x \neq 0$.

Berestycki et al. (2002) prove that

$$\lim_{\tau \downarrow 0}\widehat{\sigma}(x,\tau) = \widehat{\sigma}(x,0) \stackrel{\text{def}}{=} \left\{\int_0^1 \frac{ds}{\sigma(sx,0)}\right\}^{-1} \qquad (3.46)$$

holds in fact.

Result (3.46) establishes that for options near to expiry IV can be understood as the *harmonic mean of local volatility*. Note that – unlike the situations seen so far – the mean is taken across log-forward moneyness, i.e. in a *spatial sense* across the LVS. Berestycki et al. (2002) point out that this result relies on the particular boundary condition imposed by the *call payoff function*: $\psi(x) = (e^x - 1)^+$ (here in the inverse log-forward moneyness notation). Indeed, if it is replaced by any strictly convex function they show that $\lim_{\tau \downarrow 0}\widehat{\sigma}(x,\tau) = \sigma(x,0)$.

The authors also provide an intuitive argument for their result: consider the situation of an asset price process, the local volatility of which vanishes in some interval $[\widetilde{x}, 0]$ for $x < \widetilde{x} < 0$. Then, we get $\widehat{\sigma}(x,0) = 0$ from (3.46). Clearly, this result, which is obtained by averaging harmonically, is correct also from a probabilistic point of view, since the stock starting in x will never cross the interval and never reach the ITM region of the call. Thus the call must have a price of zero. However, an IV of zero is inconsistent with the simple (spatial) arithmetic averages.

62   3 Smile Consistent Volatility Models

For the second result, assume that local volatility is bounded away from zero and infinity and that is has the continuous limits: $\lim_{x \uparrow \infty} \sigma(x,\tau) = \sigma_+(\tau)$ and $\lim_{x \downarrow -\infty} \sigma(x,\tau) = \sigma_-(\tau)$. Then

$$\lim_{x \to \pm\infty} \widehat{\sigma}^2(x,\tau) = \frac{1}{\tau} \int_0^\tau \sigma_\pm^2(s)\, ds \,. \tag{3.47}$$

For understanding this result, note that e.g. $\widehat{\sigma}^2(+\infty, \tau) \stackrel{\text{def}}{=} \frac{1}{\tau} \int_0^\tau \sigma_+^2(s)\, ds$ has already the correct behavior by the arguments on the non-strike dependent local volatility in the previous section. To prove (3.47), Berestycki et al. (2002) construct sub- and supersolutions for any $\tau > 0$ with the required behavior at infinity and apply a comparison principle.

3.7 The Two-Times-IV-Slope Rule for Local Volatility

In our empirical demonstration of local volatility we remarked that in equity markets the slope of the local smile is approximately twice as steep as the implied smile. Derman et al. (1996b) call this empirical regularity the *two-times-IV-slope rule for local volatility*. Here, we show how this conjecture can be made more precise by using the results of the previous section.

For convenience, we reiterate the key result:

$$\widehat{\sigma}(x,0) = \left\{ \int_0^1 \frac{ds}{\sigma(sx,0)} \right\}^{-1}. \tag{3.48}$$

Consider a Taylor expansion on both sides of (3.48) in the neighborhood of $x \approx 0$. This is yields:

$$\widehat{\sigma}(0,0) + \frac{\partial \widehat{\sigma}(0,0)}{\partial x} x = \sigma(0,0) + \frac{\sigma^2(0,0)}{2} \int_0^1 \frac{\partial \sigma(0,0)}{\partial x} \frac{s\, ds}{\sigma^2(0,0)} x$$
$$= \sigma(0,0) + \frac{1}{2} \frac{\partial \sigma(0,0)}{\partial x} x \,. \tag{3.49}$$

Since $\widehat{\sigma}(0,0) = \sigma(0,0)$ by (3.48), this proves:

$$2 \frac{\partial \widehat{\sigma}(0,0)}{\partial x} = \frac{\partial \sigma(0,0)}{\partial x}, \tag{3.50}$$

i.e. the two-times-IV-slope rule holds for short-to-expiry ATM options.

We complete this section by a simulation. Suppose the local volatility smile for some close expiry date can be approximated within the interval $[-0.2, 0.2]$ by the function:

$$\sigma(x) = a(x+b)^2 + c\,, \tag{3.51}$$

3.7 The Two-Times-IV-Slope Rule for Local Volatility

where $a, b, c \in \mathbb{R}$. Computing the harmonic mean according to (3.48) yields for the IV smile

$$\widehat{\sigma}(x) = \frac{x\sqrt{ac}}{\arctan\left\{\sqrt{\frac{a}{c}}(x+b)\right\} - \arctan\left(\sqrt{\frac{a}{c}}b\right)}. \qquad (3.52)$$

In Fig. 3.3 we display the situation for $a = 0.5, b = 0.15, c = 0.3$. Note that moneyness is measured in terms of the (inverse) forward moneyness $x \stackrel{\text{def}}{=} -\ln \kappa_f$. Thus, the interval $[-0.2, 0.2]$ corresponds to $[1.22, 0.81]$ in the usual forward moneyness metric, and the smiles appear as a mirror image to Fig. 3.2. Otherwise the plots look remarkably similar. Also the two-time-IV-rule is well visible.

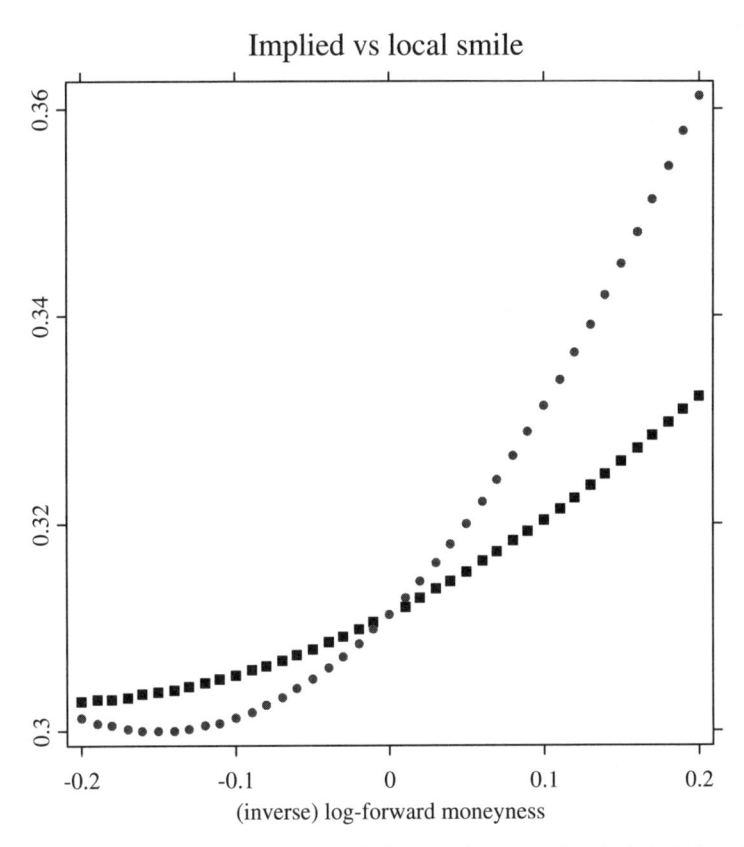

Fig. 3.3. Simulation of option implied (*squares*) versus local (*circles*) volatility smiles according to (3.51) and (3.52) for $a = 0.5, b = 0.15, c = 0.3$. Moneyness is (inverse) forward moneyness $x \stackrel{\text{def}}{=} -\ln \kappa_f$. The interval $[-0.2, 0.2]$ corresponds to $[1.22, 0.81]$ in the usual forward moneyness metric, compare Fig. 3.2

SCMsimulVLV.xpl

3.8 The K-Strike and T-Maturity Forward Risk-Adjusted Measure

As had been outlined in the introduction to this chapter, it is possible to characterize the local variance as the unconditional expectation under a K-strike and T-maturity forward risk-adjusted measure. Such a result is similar to the case of forward rates: Jamshidian (1993) prove that the forward rate can be obtained by taking the expectation of the short rate under a T-maturity forward measure.

To derive their result, Derman and Kani (1998) assume the following stochastic structure of the LVS under the objective measure P:

$$\frac{d\sigma_{K,T}^2(S_t,t)}{\sigma_{K,T}^2(S_t,t)} = \alpha_{K,T}(S_t,t)\,dt + \theta_{K,T}(S_t,t)\,dW_t^{(1)} , \qquad (3.53)$$

which we give in a simplified setting here for the sake of clarity. Originally, the authors allow for multi-factor dynamics. The process of the local variance $\left(\sigma_{K,T}^2(S_t,t)\right)_{0\leq t\leq T^*}$ is adapted to the filtration $(\mathcal{F}_t)_{0\leq t\leq T^*}$ generated by two uncorrelated Brownian motions $\left(W_t^{(0)}\right)_{0\leq t\leq T^*}$ and $\left(W_t^{(1)}\right)_{0\leq t\leq T^*}$. The drift process $\left(\alpha_{K,T}(S_t,t)\right)_{0\leq t\leq T^*}$ and the volatility process $\left(\theta_{K,T}(S_t,t)\right)_{0\leq t\leq T^*}$, which reflects the sensitivity of the LVS with respect to random shocks, are not further specified, but satisfy mild integrability and measurability conditions (see Derman and Kani (1998) for details).

In this set-up instantaneous variance is given by

$$\sigma_{S_t,t}^2(S_t,t) = \sigma_{S_t,t}^2(S_0,0) + \int_0^t \alpha_{S_t,t}(S_s,s)\,ds + \int_0^t \theta_{S_t,t}(S_s,s)\,dW_s^{(1)} , \qquad (3.54)$$

where $\sigma_{S_t,t}^2(S_0,0)$ is a known constant. Instantaneous volatility enters the asset price dynamics in the usual manner via

$$\frac{dS_t}{S_t} = \mu(S_t,t)\,dt + \sigma_{S_t,t}(S_t,t)\,dW_t^{(0)} . \qquad (3.55)$$

In this general set-up, arbitrage may be possible. To avoid arbitrage opportunities generated by (3.53) and (3.55), the drift function $\alpha_{K,T}(t,S)$ must satisfy certain conditions, similarly to those known from the Heath, Jarrow and Morton (1992) theory of interest rates. More precisely, the drift condition is given by

$$\alpha_{K,T}(S_t,t) = -\theta_{K,T}(S_t,t)\left\{\frac{1}{\phi(K,T|S_t,t)}\int_t^T\int_0^\infty \theta_{K',T'}(S_t,t)\phi(K',T'|S_t,t)\right.$$
$$\left.\times (K')^2\frac{\partial^2}{\partial(K')^2}\phi(K',T'|S_t,t)dK'dT' - \lambda^{(1)}\right\} ,$$
$$(3.56)$$

3.8 The K-Strike and T-Maturity Forward Risk-Adjusted Measure

where $\phi(K,T|S_t,t)$ denotes as usually the transition probability. The term $\lambda^{(1)}$ is the market price of *volatility* risk. Derman and Kani (1998) show the existence of a *unique* martingale measure Q if and only if the market prices of risk do not depend on K and T.

Condition (3.56) is much more involved than the classical one known from the Heath et al. (1992) theory of interest rates. This is due to the two-dimensional dependence of local volatilities on K and T. Also unlike the latter, (3.56) depends on the market price of risk and on the transition density, which render an implementation difficult. Therefore, Derman and Kani (1998) propose a discrete approximation by means of a stochastic implied tree, Sect. 3.10.2.

The dynamic evolution of local volatility under the equivalent martingale measure is given by

$$\frac{d\sigma^2_{K,T}(S_t,t)}{\sigma^2_{K,T}(S_t,t)} = \widetilde{\alpha}_{K,T}(S_t,t)\,dt + \theta_{K,T}(S_t,t)\,d\overline{W}^{(1)}_t\,, \qquad (3.57)$$

where $\widetilde{\alpha}_{K,T}(S_t,t)$ is the instantaneous drift under Q. The Brownian motion under the equivalent martingale measure is denoted by $\overline{W}^{(1)}_t$. Under this measure also the transition probability $\phi(K,T|S_t,t) = \mathsf{E}^Q\{\delta_K(S_T)|\mathcal{F}_t\}$ is a martingale. Thus, it evolves according to a SDE of the form

$$\frac{d\phi(K,T|S_t,t)}{\phi(K,T|S_t,t)} = \zeta^{(0)}_{K,T}\,d\overline{W}^{(0)}_t + \zeta^{(1)}_{K,T}\,d\overline{W}^{(1)}_t\,. \qquad (3.58)$$

The previous analysis has shown, compare (3.17), that local volatility $\sigma^2_{K,T}(S_t,t)$ obeys

$$\mathsf{E}^Q\{\sigma^2_{S_T,T}(S_T,T)\,\delta_K(S_T)|\mathcal{F}_t\} = \sigma^2_{K,T}(S_t,t)\,\mathsf{E}^Q\{\delta_K(S_T)|\mathcal{F}_t\}\,. \qquad (3.59)$$

As the transition probability on the right-hand side of (3.59), also the left-hand side of (3.59) must be a martingale. Applying Itô's lemma to the product on the right-hand side of (3.59), and collecting the drift terms arising from (3.57) and the covariation process of (3.57) and (3.58) shows that

$$\widetilde{\alpha}_{K,T}(S_t,t) + \zeta^{(1)}(S_t,t)_{K,T}\,\theta_{K,T}(S_t,t) = 0\,. \qquad (3.60)$$

Now introduce new Brownian motions $\widehat{W}^{(i)}_t = \overline{W}^{(i)}_t - \int_0^t \zeta^{(i)}_{K,T}(S_s,s)\,ds$, for $i = 0,1$. From (3.60) and (3.57) it is seen that the stochastic evolution of the local variance is given by

$$\frac{d\sigma^2_{K,T}(S_t,t)}{\sigma^2_{K,T}(S_t,t)} = \theta_{K,T}(S_t,t)\,d\widehat{W}^{(1)}_t\,, \qquad (3.61)$$

which is a martingale.

We define the new measure $Q^{(K,T)}$ via its Radon-Nikodým derivative:

$$\frac{dQ^{(K,T)}}{dQ} = \exp\left[\sum_{i=0}^{1}\left\{\int_0^T \zeta_{K,T}^{(i)}(S_s,s)\,d\overline{W}_s - \frac{1}{2}\int_0^T \left(\zeta_{K,T}^{(i)}(S_s,s)\right)^2 ds\right\}\right]. \tag{3.62}$$

This measure explicitly depends on K and T. Hence it is called the K-strike and T-maturity forward risk-adjusted measure, in analogy to the theory of interest rates. Denoting the expectation with respect to the new measure by $\mathsf{E}^{(K,T)}(\cdot)$ shows that (3.2) can be rewritten as

$$\sigma_{K,T}^2(S_t,t) \stackrel{\text{def}}{=} \mathsf{E}^Q\{\sigma_{S_T,T}^2(S_T,T)|S_T=K,\mathcal{F}_t\} = \mathsf{E}^{(K,T)}\{\sigma_{S_T,T}^2(S_T,T)|\mathcal{F}_t\}, \tag{3.63}$$

which provides the desired representation.

3.9 Model-Free (Implied) Volatility Forecasts

In a large number of studies that have been surveyed in Sect. 2.7, the quality of IV as a predictor of stock price volatility is discussed. However, it may be advantageous to resort to a volatility measure implied from options that is independent of the BS model, or at best: model-free. This goal has been achieved by Britten-Jones and Neuberger (2000). They assume that dividends and interest rates are zero. In the presence of nonzero interest rates and dividends, Britten-Jones and Neuberger (2000) interpret option and asset prices as forward prices.

Usually one is interested in comparing multi-period forecasts of volatility with volatility over several periods. To obtain the unconditional expectation of the Dupire formula (3.19), one first integrates across all strikes K:

$$\mathsf{E}^Q\{\sigma^2(S_T,T,\cdot)|\mathcal{F}_t\} = \int_0^\infty \mathsf{E}^Q\{\sigma^2(S_T,T,\cdot)|S_T=K,\mathcal{F}_t\}\phi(K,T|S_t,t)\,dK$$
$$= 2\int_0^\infty \frac{\partial C_t(K,T)}{\partial T}K^{-2}\,dK. \tag{3.64}$$

For the forecast between the two time horizons $T_1 < T_2$, integrate again with respect to time to maturity. This yields:

$$\mathsf{E}^Q\left\{\int_{T_1}^{T_2}\sigma^2(S_T,T,\cdot)|\mathcal{F}_t\right\} = 2\int_0^\infty \frac{C_t(K,T_2)-C_t(K,T_1)}{K^2}\,dK. \tag{3.65}$$

This is the unconditional expectation of the instantaneous squared volatility over a finite period $[T_1,T_2]$. Or more precisely, since the interest rate is assumed to be zero, it is the expectation of the *forward* squared volatility.

How does this forecast relate to the classical BS IV? Inserting the BS formula in (3.65) and integrating by parts reveals (after carefully examining the limits):

$$\mathsf{E}^Q\left\{\int_{T_1}^{T_2}\sigma^2(S_T,T,\cdot)|\mathcal{F}_t\right\} = 2\int_0^\infty \frac{C_t^{BS}(K,T_2)-C_t^{BS}(K,T_1)}{K^2}dK$$
$$= \sigma^2(T_2-T_1). \tag{3.66}$$

Thus, if the IVS is flat in K and T, but not necessarily a constant, squared IV, i.e. $\hat{\sigma}^2$ in our common notation, is the risk-neutral forecast as given in (3.65). There is also an intuitive argument: a lot of processes are consistent with the squared volatility forecast (3.65). Naturally, one of them is the BS deterministic (squared) volatility process. Hence, it precisely provides the forecast.

However, BS IV is a biased estimator of realized volatility, since the unbiased forecast holds only for squared volatility. This is seen from Jensen's inequality:

$$\mathsf{E}^Q\left\{\sqrt{\int_{T_1}^{T_2}\sigma^2(T,\cdot)|\mathcal{F}_t}\right\} \le \sqrt{2\int_0^\infty \frac{C_t(K,T_2)-C_t(K,T_1)}{K^2}dK} \tag{3.67}$$

Only if the IVS were a constant, i.e. if no stochastics were involved, IV would be an unbiased forecast for realized volatility. This, however, is a case of little interest.

The forecast (3.64) is a *risk-neutral* one. It will necessarily differ from the forecast under the objective measure, unless volatility risk is unpriced, and both forecasts cannot simply be compared. Nevertheless, studying the systematic deviations between realized variance and its risk-neutral forecast, would certainly contribute to our understanding of how volatility risk is priced.

3.10 Local Volatility Models

Here, we survey models and techniques to recover the LVS from observed option prices. First, deterministic implied trees are presented. They are grown either by forward induction or by backward induction. Next, trinomial trees are discussed. Stochastic implied trees are considered in Sect. 3.10.2. The section concludes with methods motivated from continuous time theory.

3.10.1 Deterministic Implied Trees

Valuation methods based on trees are working horses in option pricing. Pioneered by Cox, Ross and Rubinstein (1979) (CRR), they provide a simple framework in which pricing of path-independent and path-dependent options alike can be accomplished fast and efficiently by backward induction. Most importantly, under certain regularity conditions, they are the discrete time approximations to the diffusion

68   3 Smile Consistent Volatility Models

$$dS_t = \mu(S_t, t)\, dt + \sigma(S_t, t)\, dW_t \,, \tag{3.68}$$

where $\mu, \sigma : \mathbb{R} \times [0, T^*] \to \mathbb{R}$ are deterministic functions. As is well known, the CRR tree is the discrete time approximation of the geometric Brownian motion with constant drift and constant volatility.

In the tree framework, a given interval $[0, T]$ is divided into $j = 1, 2, \ldots, J$ equally spaced pieces of length $\Delta t = T/J$. As an approximation to (3.68) one chooses a step function starting at S_0, which jumps with a certain probability at discrete times $j, 2j, 3j, \ldots$ to one out of two (binomial tree) or out of three (trinomial tree) values in $j + 1$. Nelson and Ramaswamy (1990) discuss the conditions under which this process converges indeed to (3.68) as Δt tends to zero, and they also show how to construct a binomial approximation for a specific diffusion.

In the smile consistent *implied* lattice approaches, the tree is not specified in advance and its parameters are not inferred from a calibration to the underlying process or by an estimation from historical data of the underlying. Rather, the tree as the approximation to (3.68) is recovered from observed option data. In implied trees, the transition probabilities change from node to node, and the state space is distorted in a way which mimics the LVS reflected in the option prices. This is displayed schematically in Fig. 3.4. Thus, European options priced on this tree will correctly reproduce the IVS, and exotic options will be priced relative to them.

An implicit assumption – or from a practical point of view: a necessity – is that for any strike and any time to maturity plain vanilla option prices are available. From our discussion in Chap. 2, it is clear that this is not the case. A typical approach to resolve this problem is to smooth the IVS on the desired grid, e.g. by the smoothing techniques given in Chap. 4. Other interpolating

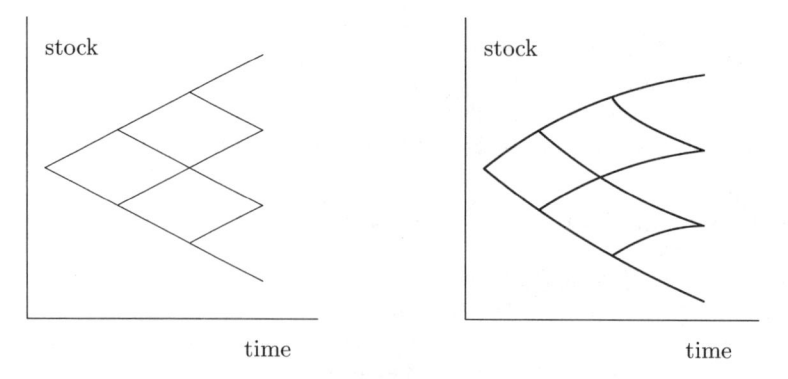

Fig. 3.4. *Left panel*: standard binomial tree, e.g. as in Cox et al. (1979). *Right panel*: implied binomial tree derived from market data, Derman and Kani (1994b)

3.10 Local Volatility Models 69

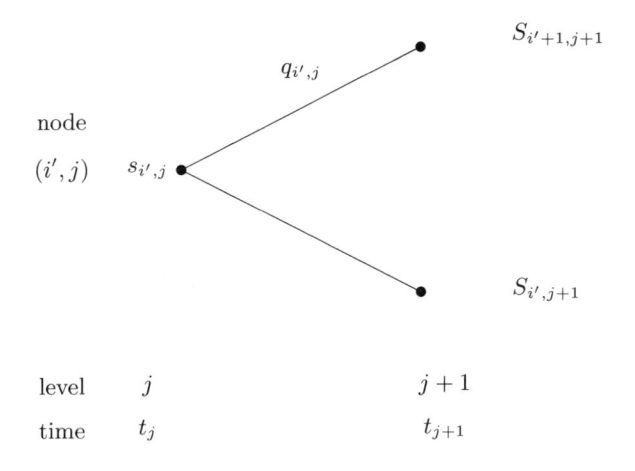

```
                                              • S_{i'+1,j+1}
                              q_{i',j}
  node
  (i', j)     s_{i',j} •

                                              • S_{i',j+1}

  level          j                               j+1
  time          t_j                              t_{j+1}
```

Fig. 3.5. Construction of the implied binomial tree from level j to level $j+1$ according to Derman and Kani (1994b) and Barle and Cakici (1998) by forward induction. $s_{i',j}$ denotes the (known) stock price at node (i',j), $S_{i'+1,j+1}$ the (unknown) stock price at node $(i'+1,j+1)$. $q_{i',j}$ is the (unknown) risk neutral transition probability from node (i',j) to node $(i'+1,j+1)$. At level j there are $i = 1, \ldots, j$ nodes (i,j).

and extrapolating techniques are a valid choice as well. The values estimated from the IVS are then inserted into the BS formula to obtain the prices of plain vanilla options at pairs of strikes and time to maturities where not available otherwise.

Derman and Kani (1994b), Barle and Cakici (1998). The principle of constructing implied binomial trees according to Derman and Kani (1994b) and Barle and Cakici (1998) is *forward induction*. The tree is (for simplicity) equally spaced with Δt and has levels $j = 1, \ldots, J$. Since the tree is recombining, there are $i = 1, \ldots, j$ nodes (i,j) at level j. The node index is running from the bottom to the top. For the presentation, suppose that the first j levels of the tree have already been implied from the option data, i.e. up to level j all stock prices $s_{i,j}$, all risk neutral transition probabilities $q_{i,j-1}$ from nodes $(i, j-1)$ to node $(i+1, j)$, and Arrow-Debreu prices $\lambda_{i,j}$ have been recovered from the option data. The Arrow-Debreu price $\lambda_{i,j}$ of node (i,j) is the price of a digital option paying one unit in this particular state and calculated as follows: ones sums over all possible paths the product of all risk neutral transition probabilities along a single path from the root of the tree to node (i,j), and discounts. In this sense the entire ensemble of the (undiscounted) Arrow-Debreu prices is the discrete version of the risk neutral transition density as introduced in Sect. 2.4.

Departing from a node (i',j) with stock price $s_{i',j}$, we consider the construction of the up-value $S_{i'+1,j+1}$ and the down-value $S_{i',j+1}$ at the nodes $(i'+1,j+1)$ and $(i',j+1)$, respectively, Fig. 3.5.

Denote by $F_{i,j} = s_{i,j}\, e^{(r-\delta)\Delta t}$ the (known) forward price maturing at time $t_{j+1} = t_j + \Delta t$, where r and δ is the interest rate and the dividend yield, respectively. Then, by risk neutrality,

$$F_{i,j} = q_{i,j} S_{i+1,j+1} + (1 - q_{i,j}) S_{i,j+1} \;. \tag{3.69}$$

There are j equations of this type, for each i one.

The second set of equations is derived from option prices, calls $C(K, t_{j+1})$ and puts $P(K, t_{j+1})$ struck at an exercise price K and expiring at t_{j+1}. Assume that

$$s_{i',j} \leq K \leq S_{i'+1,j+1} \;. \tag{3.70}$$

This choice guarantees that only the up (down) node and all nodes above (below) this node contribute to the value of the call (put) with exercise price K.

Theoretically, the prices of the call options are given from the tree by evaluating the payoff function and discounting:

$$C(K, t_{j+1}) = e^{-r\Delta t} \sum_{i=1}^{j+1} \lambda_{i,j+1}(S_{i,j+1} - K)^+ \;, \tag{3.71}$$

where

$$\lambda_{i,j+1} = \begin{cases} q_{j,j}\lambda_{j,j} & \text{for } i = j+1 \;, \\ q_{i-1,j}\lambda_{i-1,j} + (1 - q_{i,j})\lambda_{i,j} & \text{for } 2 \leq i \leq j \;, \\ (1 - q_{1,j})\lambda_{1,j} & \text{for } i = 1 \;. \end{cases} \tag{3.72}$$

In light of condition (3.70), Equation (3.71) can be written as

$$\Delta_{i'}^C = q_{i',j} \lambda_{i',j}(S_{i'+1,j+1} - K) \;, \tag{3.73}$$

where $\Delta_{i'}^C \stackrel{\text{def}}{=} C(K, t_{j+1})\, e^{r\Delta t} - \sum_{i=i'+1}^{j} \lambda_{i,j+1}(F_{i,j} - K)$. Equation (3.73) depends on the two unknown parameters $q_{i',j}$ and $S_{i'+1,j+1}$. Exploiting the risk neutrality condition (3.69), we receive from (3.73) the fundamental recursion formula for the implied binomial trees by Derman and Kani (1994b) and Barle and Cakici (1998):

$$S_{i'+1,j+1} = \frac{\Delta_{i'}^C S_{i',j+1} - \lambda_{i',j} K (F_{i',j} - S_{i',j+1})}{\Delta_{i'}^C - \lambda_{i',j}(F_{i',j} - S_{i',j+1})} \;. \tag{3.74}$$

In using (3.69) and (3.74) iteratively, one solves for $S_{i'+1,j+1}$ and $q_{i',j}$ through the upper part of the tree, if an initial $S_{i',j+1}$ is known. Indeed, there are $2j + 1$ unknown parameters in the tree at level j: $j+1$ stock prices and j transition probabilities, while the number of equations in (3.69) and (3.71) are only $2j$. This remaining degree of freedom is closed by fixing the root (the center) of the tree. If the number of nodes $j+1$ are odd, one fixes

$S_{j/2+1,j+1} = S$. Otherwise, if the number of nodes $j+1$ are even, one employs the logarithmic centering condition known from the CRR tree, i.e. one posits $S_{(j+1)/2,j+1} S_{(j+3)/2,j+1} = S^2$. Once the center is fixed the recursions (3.69) and (3.74) can be used to unfold the upper part of the tree.

Similarly, the lower part of the tree is grown from put prices. One steps down from the center, and the recursion formula (3.74) is altered to

$$S_{i',j+1} = \frac{\Delta_{i'}^P S_{i'+1,j+1} - \lambda_{i',j} K(S_{i'+1,j+1} - F_{i',j})}{\Delta_{i'}^P - \lambda_{i',j}(S_{i'+1,j+1} - F_{i',j})}. \qquad (3.75)$$

The trees by Derman and Kani (1994b) and Barle and Cakici (1998) differ in the choice of the strike prices and the centering condition. Derman and Kani (1994b) put $K = s_{i',j}$ and $S = s_{1,1}$, i.e. they fix the center of the tree at the current asset price. Barle and Cakici (1998) choose $K = F_{i',j}$ and $S = s_{1,1} e^{(r-\delta)t}$, i.e. their tree bends upward with the risk-neutral drift. They show that this choice produces a better fit to the IV smile, especially, when interest rates are very high.

Both trees are calibrated to the entire set of available option prices, both across the strike dimension and across the term structure of the IVS. However, an inherent difficulty in both trees is the fact that none of them can prevent transition probabilities from being negative. From negative transition probabilities, arbitrage possibilities ensue. Derman and Kani (1994b) avoid this by checking node by node whether $F_{i,j} < S_{i,j+1} < F_{i+1,j}$. If this condition is violated, they take a stock price that keeps the logarithmic spacing between neighboring nodes equal to the corresponding nodes at the previous level. Barle and Cakici (1998) propose to set $S_{i,j+1} = (F_{i,j} + F_{i+1,j})/2$. But even with these modifications, as the authors note, negative transition probabilities may not totally be avoided, either.

Rubinstein (1994), Jackwerth (1997). Contrary to the above approach, Rubinstein (1994) and Jackwerth (1997) construct the tree by *backward induction* beginning from a risk neutral distribution at the terminal nodes. This distribution is recovered by minimizing in a least squares sense a prior distribution, which is obtained from the binomial distribution of a standard CRR tree. The minimization is accomplished subject to the conditions of being a distribution (positivity, summability to one), and subject to correctly pricing the observed (European) option prices and the asset under the new measure. Different measures of distance do not appear to strongly affect the results of the risk neutral distribution, Jackwerth and Rubinstein (2001).

The central assumption in the tree by Rubinstein (1994) is *path independence* within the tree, i.e. the path of a downward move and an upward move is as likely as an upward move followed by a downward move. Given the known asset prices $S_{i'+1,j+1}$ and $S_{i',j+1}$ at level $j+1$ and the corresponding *nodal probabilities* $Q_{i'+1,j+1}$ and $Q_{i',j+1}$, the tree is constructed in three steps and iterated from the terminal nodes to the first one:

72   3 Smile Consistent Volatility Models

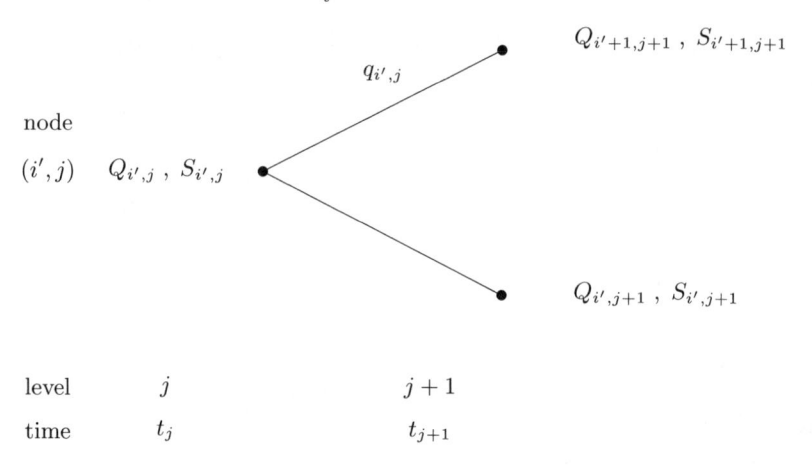

Fig. 3.6. Construction of the implied binomial tree from level $j+1$ to level j according to Rubinstein (1994) by backward induction. $S_{i',j}$ denotes the asset price at (i',j) and $Q_{i',j}$ its risk neutral *nodal* probability. $q_{i',j}$ is the (unknown) risk neutral transition probability from node (i',j) to node $(i'+1,j+1)$. Quantities at level $j+1$ are known, while those at j are unknown

$$\begin{aligned}
(1) \quad & Q_{i',j} = w(i'+1,j+1)\, Q_{i'+1,j+1} + \{1 - w(i',j+1)\}\, Q_{i',j+1}\ , \\
(2) \quad & q_{i',j} = \quad w(i'+1,j+1)\, Q_{i'+1,j+1}/Q_{i',j}\ , \\
(3) \quad & S_{i',j} = \quad e^{-(r-\delta)\Delta t}\{(1 - q_{i',j})S_{i',j+1} + q_{i',j}S_{i'+1,j+1}\}\ ,
\end{aligned} \qquad (3.76)$$

where $q_{i,j}$ denotes again the risk neutral transition probability. $w(i,j) \stackrel{\text{def}}{=} \frac{i-1}{j-1}$ is a weight function, more precisely, the fraction of the nodal probability in node (i,j) which is going down to its preceding lower node in $(i-1, j-1)$. The weight function is a consequence of the assumption of path independence and derived from the arithmetics of the CRR tree, Jackwerth (1997). Note that our notation follows Jackwerth (1997), but is adapted to observe consistency with our previous presentation: our tree has the root node $(1,1)$, which is different to both authors who start with zero.

An interesting feature of the trees implied by backward induction is that negative transition probabilities *cannot* occur by construction. This can directly be seen from (3.76). However, the crucial assumption in the tree by Rubinstein (1994) is the aforementioned property of path independence. While it facilitates the tree's construction enormously, it is also its biggest weakness: only a single maturity of options is calibrated to the tree. This may be disadvantageous when pricing exotic options, the expiry of which does not match with the maturity of the options used as inputs. This deficiency is remedied by Jackwerth (1997) in allowing for more arbitrary weight functions $w(i,j)$.

3.10 Local Volatility Models

More precisely, he proposes the piecewise function:

$$w(i,j) = \begin{cases} 2\overline{w}\frac{i-1}{j-1} & \text{for } 0 \leq \frac{i-1}{j-1} \leq \frac{1}{2} \\ -1 + 2\overline{w} + (2 - 2\overline{w})\frac{i-1}{j-1} & \text{for } \frac{1}{2} < \frac{i-1}{j-1} \leq 1 \end{cases}, \quad (3.77)$$

where $i = 1, \ldots, j$ and $0 < \overline{w} < 1$ is some value that allows $w(i,j)$ to be concave or convex in $\frac{i-1}{j-1}$. Concavity implies that a path moving down and then up, is more likely than a path moving up and afterwards down. For $\overline{w} = 0.5$, $w(i,j)$ collapses to the Rubinstein (1994) case. The choice of $\overline{w}$ can be added to the least squares problem used to recover the posterior risk neutral distribution. Jackwerth (1997) reports that a concave weight, i.e. $\overline{w} > 0.5$, explains the post-crash data (beginning from 1987) best.

Generalized binomial implied trees preserve the property that non-positive transition probabilities cannot occur, while at the same time the entire term structure of options can be employed for its construction. Furthermore, unlike the trees by Derman and Kani (1994b), Barle and Cakici (1998), and the trinomial tree by Derman et al. (1996a) to be discussed next, they are easily calibrated to non-European style options. A semi-recombining version of the trees by Rubinstein (1994) and Jackwerth (1997) is proposed by Nagot and Trommsdorff (1999).

Local Volatilities. Given an implied tree, the local volatility $\sigma_{i,j}$ at asset price level i in time step j is calculated via:

$$\mu_{i,j} = q_{i,j} R_{i+1,j+1} + (1 - q_{i,j}) R_{i,j+1},$$
$$\sigma_{i,j}^2 = q_{i,j} (R_{i+1,j+1} - \mu_{i,j})^2 + (1 - q_{i,j}) (R_{i,j+1} - \mu_{i,j})^2, \quad (3.78)$$

where $R_{i+1,j}$ denotes the return between the node $(i, j-1)$ and $(i+1, j)$ in the tree. Note that the local volatility may need to be annualized to make it comparable with IV. If we hold the horizon T of the tree fixed, and let the step size shrink to zero, the approximation tends to the local variance function of the corresponding underlying continuous time process.

Example. At this point, we illustrate the deterministic implied binomial trees using the Derman and Kani (1994b) approach. We put $r, \delta = 0$. As IVS function, we use (also displayed in Fig. 3.7):

$$\widehat{\sigma} = \frac{-0.2}{\{\ln(K/S)\}^2 + 1} + 0.3. \quad (3.79)$$

Thus, we do not model a term structure of the IVS. From this IV function, the BS option prices are computed, which are employed for growing the tree. Practically, this could be the smile function obtained from the smoothing techniques in Chap. 4.

Let's assume that $S_0 = 100$, and $T = 0.5$ years discretized in five time steps. In this case, the stock price evolution is found to be:

74   3 Smile Consistent Volatility Models

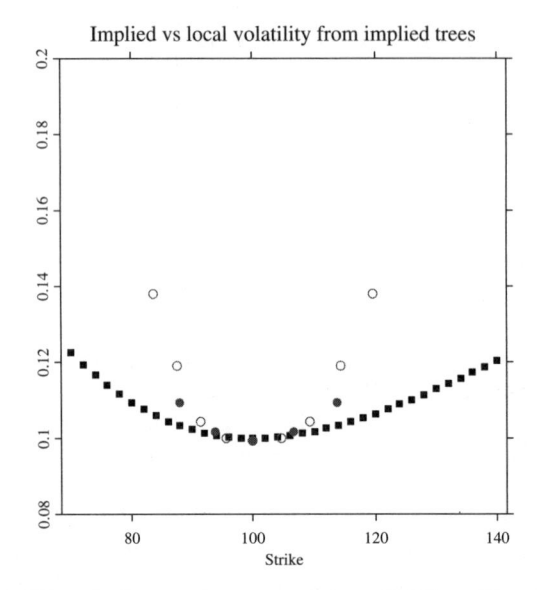

Implied vs local volatility from implied trees

Fig. 3.7. Convex IV smile (*squares*) computed from (3.79) and local (*circles*) volatility recovered from the implied binomial tree (*filled circles*) and trinomial tree (*empty circles*) SCM1bt1TTconv.xpl

```
                                                    117.9
                                           113.8
                                  110.1              110.0
                         106.6             106.5
                103.2             103.2             103.2
       100.0             100.0             100.0
                 96.9              96.9              96.9
                          93.8             93.9
                                   90.8             90.9
                                           87.8
                                                    84.8
```

The tree of the upward transition probabilities is given by:

```
                                            0.483
                                  0.486
                         0.488             0.488
                0.490             0.490
       0.492             0.492             0.492
                0.494             0.494
                         0.496             0.496
                                  0.498
                                            0.500
```

3.10 Local Volatility Models

and, finally, the tree of the Arrow-Debreu prices is:

```
                                              0.028
                                       0.057
                              0.118           0.148
                       0.241         0.242
               0.492          0.370           0.310
       1.000          0.502          0.378
               0.508          0.382           0.320
                       0.257         0.258
                              0.130           0.162
                                       0.065
                                              0.033
```

Exotic options of European style can be priced by simply multiplying the payoff function, which is evaluated at each terminal node, with the Arrow-Debreu price at this node. Since $r = 0$ we do not need to discount. For instance, for $K = 100$, the price of a digital call is the sum of the Arrow-Debreu prices for $S_T > K$: $C_{dig}(100, 1) = 0.485$. For path-dependent options, one calculates the path probabilities from the transition probabilities and iterates through the tree by backward induction.

From Equation (3.78), the tree of local volatilities is calculated as:

```
                                              0.109
                                       0.105
                              0.102           0.102
                       0.100          0.100
               0.100          0.100           0.099
       0.100          0.100          0.100
               0.100          0.100           0.102
                       0.102         0.105
                              0.105           0.109
```

In Fig. 3.7, we display the smile together with the terminal local volatilities (filled circles). It is seen that near ATM the local volatility smile is at the levels of the IV smile, but increases in either direction from ATM. This is due to the fact that the IV smile is convex. If it were monotonously decreasing, local volatility would be below IV in the right-hand side of the figure. This is seen for another example in Fig. 3.8. The two-times-IV-slope-rule is visible as well, Sect. 3.7.

76   3 Smile Consistent Volatility Models

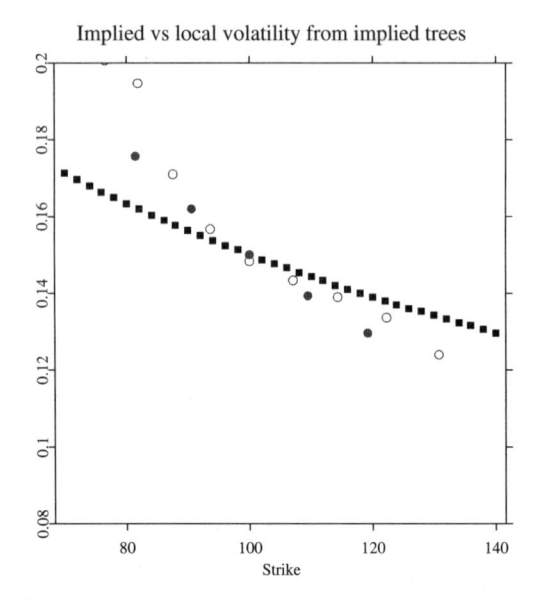

Fig. 3.8. Monotonous IV smile (*squares*) computed from $\widehat{\sigma} = -0.06\ln(K/S) + 0.15$ and local (*circles*) volatility recovered from the implied binomial tree (*filled circles*) and trinomial tree (*empty circles*) ○ SCMlbt1TTmon.xpl

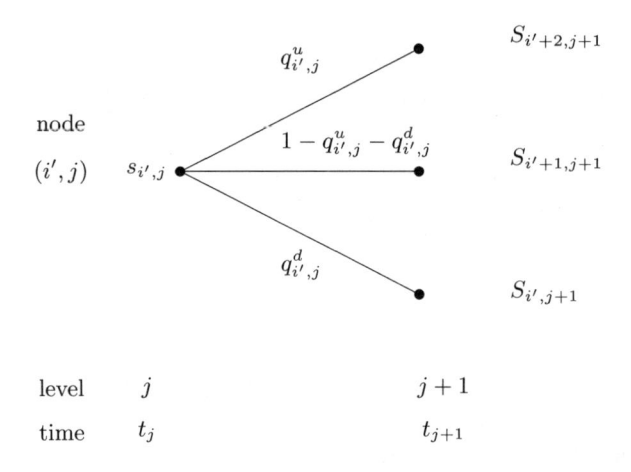

Fig. 3.9. Construction of the implied trinomial tree from level j to level $j + 1$ according to Derman et al. (1996a) by forward induction. $s_{i',j}$ denotes the (known) stock price at node (i', j), $S_{i',j+1}$ the (known, since a priori specified) stock price at node $(i', j + 1)$. $q^u_{i',j}$ is the (unknown) risk neutral transition probability from node (i', j) to the upper node $(i' + 2, j + 1)$, $q^d_{i',j}$ to $(i', j + 1)$. At level j there are $i = 1, \ldots, (2j - 1)$ nodes (i, j)

3.10 Local Volatility Models

Derman et al. (1996a). Trinomial trees provide a more flexible approximation to the state space than a binomial tree, Fig. 3.9: from each node (i,j) at a (known) stock price $s_{i,j}$, there is the possibility of an upward move to $S_{i+2,j+1}$, a downward move to $S_{i,j+1}$, and an intermediate move to $S_{i+1,j+1}$. Again we let the node index i run from the bottom to the top. As will become clear in the following, unlike the implied binomial tree which is uniquely determined (up to its trunk), the trinomial tree is underdetermined. At each node (i,j) there are five unknowns: three subsequent stock prices and two transition probabilities. Consequently, Derman et al. (1996a) propose to fix a priori the state space of the asset price evolution and to reduce the construction of the tree to backing out the transition probabilities by forward induction. We thus assume in the following that the asset price evolution has already been specified.

The trinomial tree is recovered as the binomial one. First, as in (3.69), the risk neutrality condition is

$$F_{i,j} = q_{i,j}^u S_{i+2,j+1} + (1 - q_{i,j}^u - q_{i,j}^d) S_{i+1,j+1} + q_{i,j}^d S_{i,j+1} , \qquad (3.80)$$

and the option pricing equation (3.71) for calls maturing one period later becomes:

$$C(K, t_{j+1}) = e^{-r\Delta t} \sum_{i=1}^{2j+1} \lambda_{i,j+1} (S_{i,j+1} - K, 0)^+ , \qquad (3.81)$$

where

$$\lambda_{i,j+1} = \begin{cases} \lambda_{2j-1,j} \, q_{2j-1,j}^u & \text{for } i = 2j+1 \\ \lambda_{2j-2,j} \, q_{2j-2,j}^u + \lambda_{1,j}(1 - q_{2j-1,j}^u - q_{2j-1,j}^d) & \text{for } i = 2j \\ \lambda_{i-2,j} \, q_{i-2,j}^u + \lambda_{i,j}(1 - q_{i-1,j}^u - q_{i-1,j}^d) + \lambda_{i,j} \, q_{i,j}^d & \text{for } i = 3, \ldots \\ & \ldots, 2j-1 \\ \lambda_{1,j}(1 - q_{1,j}^u - q_{1,j}^d) + \lambda_{2,j} \, q_{j,2}^d & \text{for } i = 2 \\ \lambda_{1,j} \, q_{j,1}^d & \text{for } i = 1 \end{cases} . \qquad (3.82)$$

In fixing the strike of the option at $K = S_{i'+1,j+1}$, Derman et al. (1996a) show that (3.71) together with (3.80) can be solved for the unknown transition probabilities:

$$q_{i',j}^u = \frac{e^{r\Delta t} C(S_{i'+1}, t_{j+1}) - \sum_{j=i'+1}^{2j} \lambda_{i,j}(F_{i,j} - S_{i+1,j+1})}{\lambda_{i',j}(S_{i'+2,j+1} - S_{i'+1,j+1})} , \qquad (3.83)$$

while $q_{i',j}^d$ follows immediately from (3.80). This determines the upper tree from the center, while the lower part is grown from

$$q_{i',j}^d = \frac{e^{r\Delta t} P(S_{i'+1}, t_{j+1}) - \sum_{j=0}^{i'-1} \lambda_{i,j}(S_{i+1,j+1} - F_{i,j})}{\lambda_{i',j}(S_{i'+1,j+1} - S_{i',j+1})} . \qquad (3.84)$$

Again, $q_{i',j}^u$ is given by (3.80).

78   3 Smile Consistent Volatility Models

Trinomial trees can be considered to be advantageous compared to binomial ones, since with the same number of steps, the approximation to the diffusion is finer. Thus, pricing is more accurate at a given number of steps. Furthermore they provide more flexibility, which – if judiciously handled – may help avoid negative transition probabilities as encountered in the binomial trees implied from forward induction. As a drawback, one needs to specify a priori the state space of the evolution of the asset price. Derman et al. (1996a) discuss several techniques of doing so, usually taking an equally spaced trinomial tree as starting point. From our experience, the more curved the IV function is, the easier the standard CRR tree as state space is overtaxed: more and more transition probabilities need to be overridden, which can produce unlikely local volatilities. Thus, the challenge in trinomial trees lies in an appropriate choice of the state space, which should immediately reflect the structure of the – at this point unknown! – local volatility function.

Local Volatilities. In trinomial trees, local volatilities are computed via an obvious generalization of (3.78).

Example. We illustrate the implied trinomial tree. For comparison, we put ourselves in the same situation as before. As IVS function, we use again (3.79).

For the trinomial tree, the stock price evolution fixed a priori from the CRR tree is:

```
                                          125.1
                                  119.6   119.6
                          114.4   114.4   114.4
                  109.4   109.4   109.4   109.4
          104.6   104.6   104.6   104.6   104.6
  100.0   100.0   100.0   100.0   100.0   100.0
           95.6    95.6    95.6    95.6    95.6
                   91.4    91.4    91.4    91.4
                           87.4    87.4    87.4
                                    83.6    83.6
                                            80.0
```

The tree of the upward transition probabilities is given by:

```
                                  0.466
                          0.393   0.346
                  0.296   0.276   0.267
          0.250   0.249   0.246   0.245
  0.244   0.244   0.243   0.242   0.241
          0.250   0.249   0.246   0.245
                  0.296   0.276   0.267
                          0.393   0.346
                                  0.466
```

3.10 Local Volatility Models

and the tree of the downward transition probabilities by:

```
                           0.487
                    0.411  0.362
             0.309  0.289  0.279
      0.262  0.260  0.257  0.256
0.256  0.256  0.254  0.253  0.252
      0.262  0.260  0.257  0.256
             0.309  0.289  0.279
                    0.411  0.362
                           0.487
```

Finally, the tree of the Arrow-Debreu prices is:

```
                                  0.003
                           0.007  0.010
                    0.018  0.027  0.038
             0.061  0.084  0.100  0.108
      0.244  0.241  0.229  0.215  0.202
1.000 0.500  0.378  0.316  0.278  0.251
      0.256  0.253  0.240  0.224  0.211
             0.067  0.092  0.110  0.118
                    0.021  0.031  0.044
                           0.008  0.012
                                  0.004
```

In this case, the price of the digital call is $C_{dig}(100, 1) = 0.361$. The large difference in the results of the two trees is of course due to the small number of levels used in the simulation. After increasing the levels, both prices converge to $C_{dig}(100, 1) \approx 0.40$. From (3.78) the tree of local volatilities is calculated as:

```
                           0.138
                    0.127  0.119
             0.110  0.106  0.104
      0.101  0.101  0.100  0.100
0.100 0.100  0.100  0.099  0.099
      0.101  0.101  0.100  0.100
             0.110  0.106  0.104
                    0.127  0.119
                           0.138
```

80   3 Smile Consistent Volatility Models

In Fig. 3.7, we display the smile together with the terminal local volatilities of the binomial (filled circles) and trinomial trees (empty circles). Naturally, the trinomial tree is more finely spaced.

3.10.2 Stochastic Implied Trees

Stochastic implied trees are stochastic extensions of the models discussed up to now and combine Monte Carlo and lattice approaches. They have been introduced as tractable implementations of the continuous time stochastic local volatility models as presented in Sect. 3.8. Let $(\Omega, \mathcal{F}, Q)$ be a probability space with some martingale measure Q, which is equipped with the filtration $(\mathcal{F}_t)_{0 \leq t \leq T}$. The key idea is to stochastically perturb the LVS observed for a given set of option prices. While the asset price, which is adapted to $(\mathcal{F}_t)_{0 \leq t \leq T}$, moves randomly from node to node through the state space, local transition probabilities between the nodes vary as well, thereby reflecting the stochastic perturbations in local volatilities.

Derman and Kani (1998). Starting point in Derman and Kani (1998) is the trinomial tree introduced by Derman et al. (1996a) which is calibrated to the set of observed option prices. Next local volatilities are perturbed by the discretized SDE

$$\Delta \sigma_{m,n}^2(i,j) = \sigma_{m,n}^2(i,j) \left\{ \widetilde{\alpha}_{m,n}(i,j) \Delta t_j + \theta \Delta \overline{W}^{(1)} \right\}, \quad (3.85)$$

where the pair (i,j) denote the node (S_i, t_j) in the tree, while (m,n) denote all future nodes in the tree. This equation is meant to discretize the SDE in (3.57).

The volatility parameter θ is chosen in advance, e.g. via the principle component analysis (PCA) presented in Chap. 5, while the drift coefficients $\widetilde{\alpha}_{m,n}(i,j)$ are obtained from the no-arbitrage requirement that the total probability $Q_{m,n}(i,j)$ of arriving at the future node (n,m) from the fixed initial node (i,j) must be jointly martingales for all future nodes (n,m). Next a random vector denoted by $(\Delta \overline{W}^{(0)}, \Delta \overline{W}^{(1)})^\top$ is drawn. The first entry is used to determine a new level of the underlying asset given by the three subsequent nodes in the tree. The second one is directly inserted into (3.85) to arrive at a new location for the entire volatility surface $\sigma_{m,n}^2(i, j+1)$. In the following, all steps described are repeated for each node (i,j) in the tree. After each draw of the random sample, the new drift coefficients are calculated from the conditions on $Q_{m,n}(i,j)$, and so on. Thus, one generates many sample paths through the tree as random realizations of arbitrage-free dynamics.

In specification (3.85), $\overline{W}^{(1)}$ is interpreted as a proportional shift to all local volatilities. This corresponds to the main source of noise in the IVS as shall be seen in Chap. 5. It is natural to assume that this holds also for the LVS. Of course, multi-factor, node-dependent dynamics for the LVS could be specified as well. Also parametric choices of the eigenfunctions recovered by the functional PCA methods in Sect. 5.3 could be used. They could be chosen in a way to model slope and twist shocks in the surface.

Stochastic implied trees are a flexible framework for option pricing and hedging, since they also comprise non-Markovian volatility processes. However, the calculation of the drift-parameters becomes increasingly involved, and the tree must be recalculated in each single simulation step, which is computationally very demanding. Also, since the state space remains fixed from the beginning, negative transition probabilities may occur when the volatilities become very large. They need to be overwritten manually. Derman and Kani (1998) report for their simulations that this occurred in less than 3% of all paths simulated.

Britten-Jones and Neuberger (2000). In following the work by Derman and Kani (1998), Britten-Jones and Neuberger (2000) propose a trinomial implied tree that allows for stochastic volatility. Unlike the former approach, their setting is Markovian, but for this reason also much simpler. As usual in the trinomial tree framework, they start on a discrete time interval $h = \Delta t$ by fixing the state space of the asset price evolution under the risk neutral measure Q. The state space is chosen to be a finite geometric series (without loss of generality):

$$\mathcal{K} = \{K|\ K = S_0 u^j\ ,\ j = 0 \pm 1, \pm 2, \ldots, \pm T/h\}\ , \qquad (3.86)$$

where $u > 0$. Additionally, they require that if $|j - k| > 1$, then $Q(S_{t+h} = S_0 u^k | S_t = S_0 u^j) = 0$ with $j, k = 0 \pm 1, \pm 2, \ldots, \pm T/h$. The latter assumption can be thought of as a continuity assumption. As data input they require that a complete set of European calls $C(K, t)$ for all expirations $t \in \mathcal{T} = \{0, h, 2h, \ldots, T\}$ and strikes $K \in \mathcal{K}$ be given.

Define the quantities

$$\Pi(K, t) \stackrel{\text{def}}{=} \frac{C(Ku, t) - (1+u)C(K, t) + uC(K/u, t)}{K(u-1)}\ , \qquad (3.87)$$

$$\Lambda(K, t) \stackrel{\text{def}}{=} \frac{C(K, t+h) - C(K, t)}{C(Ku, t) - (1+u)C(K, t) + uC(K/u, t)}\ . \qquad (3.88)$$

Note that $\Pi(K, t)$ is the cost of a butterfly spread paying one euro, if $S_t = K$, and zero otherwise. Thus it is the Arrow-Debreu security in this framework.

Assuming that the asset price process adapted to the filtration $(\mathcal{F}_t)_{0 \leq t \leq T}$ is a martingale with respect to the risk neutral measure Q, they show:

$$Q(S_t = K|\mathcal{F}_0) = \Pi(K, t) \text{ for all } t \in \mathcal{T}, K \in \mathcal{K}\ , \qquad (3.89)$$

and

$$Q(S_{t+h} = K'|S_t = K, \mathcal{F}_0) = \begin{cases} \Lambda(K, t) & \text{if } K' = Ku \\ 1 - (1+u)\Lambda(K, t) & \text{if } K' = K \\ u\Lambda(K, t) & \text{if } K' = K/u \end{cases}\ . \qquad (3.90)$$

In (3.89) it is seen that the probability of the asset arriving at any price level in the tree on a future date $t \in \mathcal{T}$ is fully determined by an initial set of option prices. However, it is also obvious from (3.89) and (3.90) that this does *not* determine the probability of a specific price path, since the conditioning information in (3.90) is neither $\mathcal{F}_t$ nor the price history up to t. Thus prices of exotic options are not unique. The probability of a price path would be determined if (and only if) the price process were Markovian, i.e. if

$$Q(S_t = K | \mathcal{F}_{t-1}) = Q(S_t = K | S_{t-1}) \text{ for all } t. \tag{3.91}$$

This would be the case if the volatility were fully deterministic in S and t. Thus, under the assumption of a deterministic volatility this approach can be used to recover the complete price process from option prices.

Under stochastic volatility, however, all risk-neutral processes consistent with the initial option prices share that the expectation of the squared returns is given by:

$$\mathsf{E}^Q \left\{ \left(\frac{S_{t+h} - S_t}{S_t} \right)^2 \Big| S_t = K \right\} = \Lambda(K, t) \frac{(u-1)^2(u+1)}{u}. \tag{3.92}$$

This is a necessary and sufficient condition, and the discrete-time counterpart of the Dupire formula (3.19) in this tree.

To implement their stochastic volatility framework Britten-Jones and Neuberger (2000) assume the existence of a time-homogenous Markov chain Z that affects the one-step transition probabilities, i.e. the local volatilities in the tree. The chain $Z \in \{1, 2, \ldots, N\}$ takes values on a set of integers with the transition matrix defined by its elements $\mathbf{Q} = (q_{m,n})$, where $q_{m,n} = Q(Z_{t+h} = m | Z_t = n, \mathcal{F}_t)$. The transition probabilities are chosen independently and depend on the specific volatility process to be modelled.

Define $\Pi(K, t, z) \stackrel{\text{def}}{=} Q(S_t = K \text{ and } Z_t = z | \mathcal{F}_0)$ and $\Lambda(K, t, z) \stackrel{\text{def}}{=} Q(S_t = Ku | S_t = K \text{ and } Z_t = z, \mathcal{F}_0)$. The authors show that – in order to be consistent with the initial set of option prices – $\Lambda(K, t, z)$ must satisfy:

$$\Lambda(K, t) \Pi(K, t) = \sum_{n=1}^{N} \Lambda(K, t, n) \Pi(K, t, n). \tag{3.93}$$

The left-hand side of (3.93) is extracted from the option data. In order to identify the right-hand side they put $\Lambda(K, t, z) = \widetilde{q}(K, t) \, v(z)$, where $v(z)$ is an exogenously chosen volatility function depending on the state z, and $\widetilde{q}(K, t)$ a multiplicative, node-dependent drift adjustment.

If all $\Pi(K, t, z)$ and $\widetilde{q}(K, t)$ are known for all prices K and volatility states z up to t, forward induction of the tree is done via the following two steps: first imply

3.10 Local Volatility Models

$$\Pi(K, t+h, z) = \sum_{n=1}^{N} q_{z,n} \Big[\Lambda(K/u, t, n) \Pi(K/u, t, n) \\ + u\Lambda(Ku, t, n) \Pi(Ku, t, n) \\ + \{1 - (1+u)\Lambda(K, t, n)\} \Pi(K, t, n) \Big]. \quad (3.94)$$

Second calculate the adjustments from

$$\widetilde{q}(K, t+h) = \frac{\lambda(K, t+1) \Pi(K, t+h)}{\sum_n v(n) \Pi(K, t+h, n)}, \quad \text{for} \quad K = S_0 u^j, \ |j| \le t/h. \quad (3.95)$$

The first step (3.94) follows a discrete version of the forward Kolmogorov equation, in that the probability of a time-dependent state event is expressed as the sum of the products of the preceding events and the one-step transition probabilities. Equation (3.95) is obtained from (3.93).

Pricing works via backward valuation. Let $V(K, t, z)$ be the value of an option depending on level K and volatility state z. It has the terminal payoff $V(K, T, z)$. By the backward iteration

$$V(K, t-h, z) = \sum_{n=1}^{N} q_{z,n} \Big[\Lambda(K/u, t-h, n) V(Ku, t, n) \\ + u\Lambda(Ku, t-h, n) V(K/u, t, n) \\ + \{1 - (1+u)\Lambda(K, t-h, n)\} V(K, t, n) \Big], \quad (3.96)$$

the price of the option is computed. Any contingent claim can be valued using the lattice but the prices depend on the volatility process chosen.

The approach by Britten-Jones and Neuberger (2000) is an elegant, and fast methodology for valuing options under stochastic volatility. It allows for a wide range of volatility specifications including mean-reversion, GARCH, or regime-switching models. Rossi (2002) investigates the ability of this model to capture the smile dynamics among alternative volatility specifications.

Another recent advance in stochastic local volatility model is an approach by Alexander et al. (2003): they model the local volatility function by a stochastic mixture of local variances derived from a small number of base processes. From this point of view they extend the work by Brigo and Mercurio (2001) discussed in Sect. 3.10.3. Alexander et al. (2003) report that the model captures the patterns of the IVS both for short and long time to maturities very well. Overall, stochastic local volatility models appear to be a fruitful line of research. Their empirical performance in hedging and pricing, for instance along the lines of Dumas et al. (1998) and Rosenberg (2000), remains to be investigated more deeply.

3.10.3 Reconstructing the LVS

Parametric Approaches

In this section, we survey approaches that aim at identifying the LVS as a continuous function. In parametric approaches a functional form of the local volatility is chosen and calibrated to the market data. As pioneering work for alternative volatility specifications one may consider the constant elasticity of variance model due to Cox and Ross (1976). In this model instantaneous volatility is specified as

$$\sigma(S_t, t) = \sigma S_t^{\alpha - 1} , \qquad (3.97)$$

with constants $\sigma, \alpha > 0$. Since volatility is a deterministic function in S_t, the LVS is:

$$\sqrt{\sigma_{K,T}^2(S_t, t)} = \sigma(K, T) = \sigma K^{(\alpha - 1)} . \qquad (3.98)$$

For $\alpha = 1$ we obtain the BS case. When $\alpha < 1$, the volatility increases as the stock price decreases. This corresponds to a transition probability function with heavy left tail and less heavy right tail. Consequently, this model produces a downward sloping IV smile.

Another type of models that received recent attention are the quadratic volatility models, Ingersoll (1997) and Rady (1997). Note however that for this class of models, the term volatility refers to the function $\tilde{\sigma}(S_t, t)$ in the SDE of the form:

$$dS_t = \mu(S_t, t)\, dt + \tilde{\sigma}(S_t, t)\, dW_t , \qquad (3.99)$$

which is unlike our terminology. Typically $\tilde{\sigma}(S_t, t) \stackrel{\text{def}}{=} \gamma(t)\, p(S_t)$ for a strictly positive and bounded function γ and a quadratic polynomial $p(x) = a + bx + cx^2$. Zühlsdorff (2002) shows existence and uniqueness of the solution to (3.99) and also discusses option pricing, when p has no, one and two real roots. According to his simulations this model is perfectly able to mimic the smile patterns one usually observes in the markets. An empirical application with bounded polynomials up to order two in asset prices and time to maturity is given by Dumas et al. (1998).

Other more flexible specifications have been proposed: Brown and Randall (1999) use sums of hyperbolic trigonometric functions designed to capture the term structure, smile and skew effects in the surface. Piecewise quadratic and cubic splines are employed by Beaglehole and Chebanier (2002) and Coleman et al. (1999), respectively. McIntyre (2001) approximates the LVS with Hermite polynomials.

The general advantage of these approaches appears to be that the estimated LVS does not exhibit excessive spikes as fully nonparametric calibrations are prone to unless strongly regularized. However, the parametric calibration problem can be underdetermined given the small number of observed market prices and the large number of parameters. Thus, the optimal parameters may not be uniquely identifiable, which may cause instability for instance in the computation of value at risk measures, Bouchouev and Isakov (1999).

Mixture Diffusions

A very flexible parametric, yet parsimonious modeling strategy based on mixture diffusions was introduced by Brigo and Mercurio (2001). Let the dynamics of the asset price under the risk-neutral measure Q be given by:

$$\frac{dS_t}{S_t} = (r - \delta)\,dt + \sigma(S_t, t)\,d\overline{W}_t^{(0)}\,, \qquad (3.100)$$

where $\sigma(S_t, t)$ is a deterministic function satisfying the linear-growth condition spelled out in Appendix B to guarantee a unique solution to this SDE. Furthermore, we are given N diffusions

$$dS_t^{(i)} = (r - \delta)S_t^{(i)}\,dt + \theta_i\big(S_t^{(i)}, t\big)\,d\overline{W}_t^{(0)},\ \ i = 1,\ldots, N\,, \qquad (3.101)$$

with common initial value S_0. The volatility functions $\theta_i(\cdot)$ satisfy similar growth-conditions. Denote by $\phi_i(K, T|S_t, t)$ the risk neutral transition density of these processes. The task is to identify the volatility function of (3.100) such that the risk neutral transition density satisfies:

$$\phi(K, T|S_t, t) = \sum_{i=1}^{N} \lambda_i \phi_i(K, T|S_t, t)\,, \qquad (3.102)$$

where $\lambda_i \geq 0$ and $\sum_{i=1}^{N} \lambda_i = 1$.

As shown in Brigo and Mercurio (2001), the solution is found by inserting the candidate solution (3.102) into the Fokker-Planck equation (see Appendix B) and solving for the variance function by integrating twice. The solution is given by:

$$\sigma^2(S_t, t) = \frac{\sum_{i=1}^{N} \lambda_i\,\theta_i^2(S_t, t)\,\phi_i(\cdot)}{\sum_{i=1}^{N} \lambda_i\,S_t^2\,\phi_i(\cdot)}\,. \qquad (3.103)$$

In the special case, where $\theta_i(S_t, t) \stackrel{\text{def}}{=} \theta_i(t) S_t$, the variance can be written as a weighted average of the individual variance functions:

$$\sigma^2(S, t) = \sum_{i=1}^{N} \widetilde{\lambda}_i\,\theta_i^2(t)\,, \qquad (3.104)$$

where $\widetilde{\lambda}_i \stackrel{\text{def}}{=} \lambda_i \phi_i(\cdot)/\phi(\cdot)$.

Hence the asset price process satisfies:

$$\frac{dS_t}{S_t} = (r - \delta)\,dt + \sqrt{\sum_{i=1}^{N} \widetilde{\lambda}_i\,\theta_i^2(t)}\,d\overline{W}_t^{(0)}\,. \qquad (3.105)$$

Brigo and Mercurio (2001) point out that the conditions for existence and uniqueness of a strong solution to (3.105) must be given case by case for different specifications of the base transition densities $\phi_i(\cdot)$. Brigo et al. (2002) and Brigo and Mercurio (2002) analyze the cases of mixtures of normals, log-normals and sine-hyperbolic processes.

The elegance of this approach becomes apparent in option pricing, especially when there are analytical pricing formulae for the base transition densities. Due to linearity of the integration and derivative operators, the price H_t of an option is given by

$$H_t = e^{-r\tau} \mathsf{E}^Q\{\psi(S_T)|\mathcal{F}_t\}$$
$$= e^{-r\tau} \int_0^\infty \psi(S_T) \sum_{i=1}^N \phi_i(K,T|S_t,t)\, dK$$
$$= \sum_{i=1}^N \lambda_i H_t^{(i)}, \tag{3.106}$$

where ψ is some payoff function and $H_t^{(i)}$ denotes the corresponding option prices of the base processes. Also all greeks of H_t are convex sums of the base option greeks. In the special case of log-normal mixtures, option prices are weighted sums of the BS prices of the options in the base processes which makes the computation of prices particularly easy.

The approach is beautiful, since it provides a close link between the local volatility and the risk neutral transition density. In the aforementioned approaches it is difficult, if not impossible, to determine the risk neutral transition density from its specific parameterization at hand. However, as was seen, this is desirable as it can make the computation of hedge ratios and prices more straightforward, especially, when closed-form solutions are available.

Nonparametric Methods

Alternative to the approaches above, another strand of literature directly aims at recovering the full LVS directly from a set of observed option prices. After estimating the LVS, it is implemented into pricing algorithms, e.g. finite difference schemes to solve the generalized BS PDE, Randall and Tavella (2000). Formally, reconstructing the LVS from option prices is an inverse problem. Since the number of parameters for the calibration of the volatility surface largely outnumber the number of observations, which are typically very small, Sect. 2.5, the problem is ill-posed in general. A review on this literature is given by Bouchouev and Isakov (1999). They distinguish three main approaches of numerical methods: optimization based algorithms, extra- and interpolation schemes and iterative procedures.

Optimization based algorithms recover the LVS directly from the generalized BS PDE (2.67) or from the dual PDE (3.19) by optimizing some cost

3.10 Local Volatility Models

functional subject to the appropriate boundary conditions. Due to the ill-posedness of the problem, small perturbations of the input data tend to result in very different solutions of the minimizing functional. In order to stabilize the computation, regularization methods for calibration are implemented. In the *Tikhonov regularization*, one adds a smoothing device which insures that the optimization problem has a unique solution under some goodness-of-fit measure. For instance, Lagnado and Osher (1997) minimize the L^2-norm of the gradient of the LVS such that the squared difference of the theoretical and the observed prices is as close as possible to zero. The 'closeness' is steered by a parameter to be chosen by the user. In each step the variational derivatives are calculated for each point on a finite difference grid in a steepest descent minimization. Berestycki et al. (2002) formulate the regularized cost functional based on their asymptotic results reported in Sect. 3.6. Alternative approaches using Tikhonov regularization are Jackson et al. (1998), Bodurtha and Jermakyan (1999), Coleman et al. (1999), and Bodurtha (2000). As an alternative means of regularization, Avellaneda et al. (1997) minimize the *relative-entropy* distance to a prior distribution. They solve a constrained optimal control problem for a Bellman parabolic equation. An optimal control framework is also chosen by Jiang and Tao (2001) and Jiang et al. (2003) to determine the LVS.

Particularly simple methods are extra- and interpolation techniques. They discretize the dual PDE (3.19) and extra- and interpolate the data for all strikes and maturities. This is most conveniently achieved in the IV representation of the LVS as derived in Equation (3.36). As extra- and interpolation techniques Andersen and Brotherton-Ratcliffe (1997) and Dempster and Richards (2000) employ cubic splines. Typically the splines are fitted first across strikes, only, and a second set of splines across maturities. The relevant derivatives are computed and inserted into (3.36). The evident disadvantage of this approach is that the smoothness of the derivatives is guaranteed only in the strike direction. In our computations of the LVS, we overcome this point in employing the second order local polynomial estimator to estimate the IVS. In this case, all derivatives are natural byproducts of the estimation. A possible drawback of extra- and interpolation methods is that they are sensitive and unstable. A particular challenge is the extrapolation of the IVS into areas where no IVs are observed. Theoretical results by Lee (2003) suggest that the smile function should be extended in log-moneyness as a square root multiplied with a constant reflecting the number of moments assumed to exist in the underlying risk neutral distribution, Sect. 2.6.2. The difficulty of extra- and interpolation methods is that there does not appear to be a way to guarantee that standard arbitrage bounds are not violated, or that the local variance remains positive and finite.

Finally, given the small number of observations, Bouchouev and Isakov (1999) suggest iterative procedures to reconstruct the LVS. In their first approach they employ an analytic approximation for the solution to the generalized BS PDE. This leads to an integral equation for the LVS that is discretized

88   3 Smile Consistent Volatility Models

at the points where the option data are available. A resulting system of non-linear equations is solved, where the values of the LVS are recovered as BS IVs from adjusted option prices. His second method iteratively exploits the fundamental solution to the generalized BS PDE (2.67).

To our knowledge, little is known how the different algorithms compare among each other. Especially, a comprehensive appraisal of the different approaches in terms of stability, computational costs, and proneness to errors remains to be done.

3.11 Excellent Fit, but...: the Delta Problem

As has been pointed out throughout this work, the decisive virtue of smile consistent models, local volatility models in particular, is that they completely reproduce or reprice the market, thereby allowing to price plain vanilla options and exotic options alike with the same model. This is simply by construction, and theoretically appealing, since a lot of types of exotic options can be hedged via static approaches, Derman et al. (1995), Carr et al. (1998) and Andersen et al. (2002). However, since the conditions under which static hedging works, are typically not met on real markets, in practice one often hedges dynamically. Dynamic hedging depends on the accuracy to which the greeks describe the price dynamics to first or second order. However, this is exactly where local volatility models have been put severely under fire in a article by Hagan et al. (2002). The authors focus their criticism on the delta computed from local volatility models.

To illustrate their main argument, they consider the special case where local volatility is a function of the form:

$$\sigma(S_t, t) = \sigma(S_t), \qquad (3.107)$$

where σ is deterministic. By singular perturbation techniques, Hagan and Woodward (1999) and Hagan et al. (2002) show that todays IV function $\hat{\sigma}_0(S_0, K)$ is related to local volatility to leading order by:

$$\hat{\sigma}_0(S_0, K) = \sigma\left\{\frac{1}{2}(S_0 + K)\right\}\left[1 + \frac{1}{24}\frac{\sigma''\left\{\frac{1}{2}(S_0 + K)\right\}}{\sigma\left\{\frac{1}{2}(S_0 + K)\right\}}(S - K)^2 \ldots\right], \qquad (3.108)$$

where σ'' denotes the second derivative of the local volatility function with respect to S. According to Hagan et al. (2002) the first term in (3.108) accounts already for 99% of the IV function. Therefore, one can safely pretend that

$$\hat{\sigma}_0(S_0, K) \approx \sigma\left\{\frac{1}{2}(S_0 + K)\right\}, \qquad (3.109)$$

which also uncovers the two-times-IV-slope rule, Sect. 3.7. Thus, for a given IV function the fitted local volatility must satisfy (3.109), or equivalently, at a strike $K = 2S - S_0$, we find that

3.11 Excellent Fit, but...: the Delta Problem

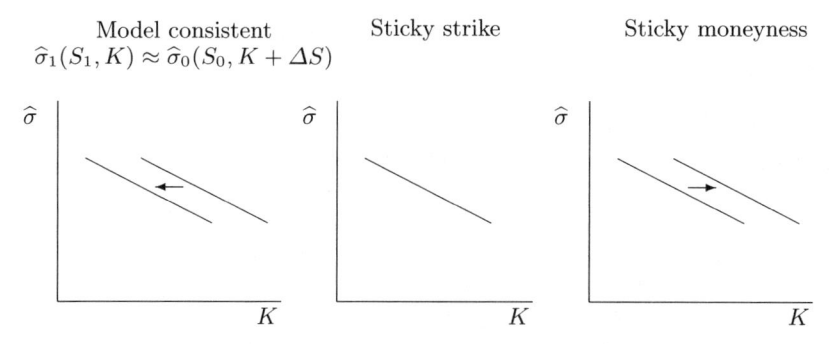

Fig. 3.10. Alternative IV smile dynamics assuming an upward shift of the asset price. *Left panel*: dynamics of the IV smile implied from (deterministic) local volatility models. *Central panel*: sticky-strike assumption. *Right panel*: sticky-moneyness assumption

$$\widehat{\sigma}_0(S_0, 2S - S_0) \approx \sigma\left\{\frac{1}{2}(S_0 + 2S - S_0)\right\} = \sigma(S) . \qquad (3.110)$$

Putted in words, this means that the local volatility at some point S corresponds (approximately) to the IV function at the strike $K = 2S - S_0$.

Suppose now that the current spot value S_0 changes by ΔS to S_1. The decisive point to remember is that the local volatility function remains the same, it is simply evaluated at the new spot level $S_1 = S_0 + \Delta S$. Therefore, reading Equation (3.109) from left to right and from right to left shows that the new IV smile is related to the previous one by:

$$\widehat{\sigma}_1(S_1, K) \approx \sigma\left\{\frac{1}{2}(S_0 + \Delta S + K)\right\} \approx \widehat{\sigma}_0(S_0, K + \Delta S) . \qquad (3.111)$$

Thus, as the spot moves up, the smile shifts to the left, and vice versa. This behavior, however, is against the common market experience, Fig. 3.10: instead, the smile is expected to remain constant at the strikes (*sticky-strike*-assumption), i.e. the smile function for a given K does not change, or to shift with the spot (*sticky-moneyness*-assumption), i.e. the smile stays constant measured in terms of moneyness, Derman (1999).

Now, consider the delta in the local volatility model, which is simply given the BS delta and a vega correction, compare with Sect. 2.9:

$$\frac{\partial \widetilde{C}_t}{\partial S} = \frac{\partial C_t^{BS}}{\partial S} + \frac{\partial C_t^{BS}}{\partial \widehat{\sigma}} \frac{\partial \widehat{\sigma}}{\partial S} . \qquad (3.112)$$

Since the local volatility model predicts that the smile moves left when the spot moves up and vice versa, which is opposite to common market behavior, Hagan et al. (2002) conclude that the local volatility delta is wrong or at best very misleading.

In practice this problem is met by recalibration of the model. Instead of reading the delta from the finite difference scheme, which yields the model-implied delta, one shifts the spot and computes the delta via a finite difference quotient. In shifting the spot, one imposes the IV smile dynamics that are considered as appropriate, i.e. one recomputes the new option prices either at the same smile (sticky-strike), or at a smile function shifted with the spot (sticky-moneyness). This practice, however, has led to a whole delta menu and a fierce debate on which is the best: the model-implied local volatility delta, the sticky-strike or BS delta, and the sticky-moneyness delta.

From a theoretical perspective, the answer can be given case by case depending on the prevailing market regime, Derman (1999) and more recently Crépey (2004), but practically the question appears to be unsolved. In simulating alternative asset price dynamics, McIntyre (2001) finds that the local volatility model is not delivering robust delta hedges when the true model is a jump-diffusion, but fairly accurate ones in a pure stochastic volatility setting. In hedging exercises with real data, Dumas et al. (1998) prefer the sticky-strike delta to the local volatility variant, whereas Coleman et al. (2001) and Vähämaa (2004) find opposite evidence. Clearly, the contradicting results can be due to the fact that the 'right' delta depends on the current market regime, and a final answer cannot be given, or must be sought in stochastic local volatility settings, Alexander and Nogueira (2004).

Clearly, the delta discussion extends also to other higher order greeks involving a spot derivative, in particular gamma and vanna, but the literature appears to be silent on this topic. The difficulty is that higher order greeks are prone to numerical errors making an analysis very cumbersome. But still, since the local volatility models are frequently used for options with non-convex payoff profiles, such as barrier options, this discussion is of vital importance, and needs to be addressed in the future.

Aside from the delta problem, another unsatisfying feature of LV models is that they predict flat future smiles: the since the IVS flattens out for longer time horizons, so does the LVS, compare Fig. 3.1. Therefore, implicitly the model predicts flat future smiles, which is typically not what one expects. Therefore, options that start in the long dated future, such as forward start options and cliquet structures, will be priced incorrectly, as their prices are computed under the assumption of a flat (forward) IVS at their starting date. These types of exotics need to be priced with stochastic LV model, stochastic volatility or jump diffusion models that do not suffer from this drawback, Kruse (2003).

3.12 Stochastic IV Models

Stochastic IV models follow a different strategy than the local volatility and the classical stochastic volatility models: the idea is not to introduce the stochastic setting via the instantaneous volatility function, but through a stochastic IV process. Like deterministic local volatility models, they allow for a preference-free option valuation, since markets are complete owing to the fact that volatility is tradable through options, usually plain vanilla options of European style. Stochastic IV models were developed by Ledoit and Santa-Clara (1998) and Schönbucher (1999), and have recently been more deeply analyzed by Brace et al. (2001), Amerio et al. (2003) and Daglish et al. (2003).

The (somewhat simplified) model set-up is as follows: for a fixed time interval $[0, T^*]$, we consider a probability space $(\Omega, \mathcal{F}, Q)$, where Q is the (unique) martingale measure in the economy. We define two Brownian motions $\left(\overline{W}_t^{(0)}\right)_{0 \leq t \leq T^*}$ and $\left(\overline{W}_t^{(1)}\right)_{0 \leq t \leq T^*}$ on this space. Without loss of generality they are assumed to be uncorrelated. The space is equipped with a filtration $(\mathcal{F}_t)_{0 \leq t \leq T^*}$. As tradable assets, we have the underlying asset S_t paying a constant dividend yield δ, a riskless investment with constant interest rate r, and a European call option $C(S_t, t, K, T)$.

Under the measure Q the asset price dynamics are governed by the SDE

$$\frac{dS_t}{S_t} = (r - \delta)\, dt + \sigma(S_t, t, \widehat{\sigma}_t)\, d\overline{W}_t^{(0)}, \qquad (3.113)$$

where $\left(\sigma(S_t, t, \widehat{\sigma}_t)\right)_{0 \leq t \leq T^*}$ is some $(\mathcal{F}_t)_{0 \leq t \leq T^*}$-adapted stochastic process. It will be seen that it is driven by the stochastic IV process which follows

$$\frac{d\widehat{\sigma}_t(K,T)}{\widehat{\sigma}_t(K,T)} = \alpha(\widehat{\sigma}_t, t, S_t)\, dt + \theta_0(\widehat{\sigma}_t, t, S_t)\, d\overline{W}_t^{(0)} + \theta_1(\widehat{\sigma}_t, t, S_t)\, d\overline{W}_t^{(1)}, \quad (3.114)$$

where $\left(\alpha(\widehat{\sigma}_t, t, S_t)\right)_{0 \leq t \leq T^*}$ and $\left(\theta_i(\widehat{\sigma}_t, t, S_t)\right)_{0 \leq t \leq T^*}$ are predictable stochastic processes. The explicit dependence on $(\widehat{\sigma}_t, t, S_t)$ is dropped in the following for the sake of clarity. Also we will write $\widehat{\sigma}_t$ only, but the dependence of IV on K and T should be borne in mind. Finally, all diffusion parameters are assumed to satisfy the regularity assumptions such that unique strong solutions exist, see appendix Chap. B. The option is priced using the BS formula together with the current realization of the IV process $\widehat{\sigma}_t$.

A first set of restrictions on the drift of the IV process insures that no-arbitrage opportunities exist. They are derived as follows: by Itô's lemma the dynamics of the call are given by:

$$\begin{aligned}
dC_t = {} & \frac{\partial C_t}{\partial t}\, dt + \frac{\partial C_t}{\partial S}\, dS_t + \frac{1}{2}\sigma^2(S_t, t, \widehat{\sigma}_t) S_t^2 \frac{\partial^2 C_t}{\partial S^2}\, dt \\
& + \frac{\partial C_t}{\partial \widehat{\sigma}}\, d\widehat{\sigma}_t + \frac{1}{2}\frac{\partial^2 C_t}{\partial \widehat{\sigma} \partial \widehat{\sigma}}\, d\langle \widehat{\sigma} \rangle_t + \frac{\partial^2 C_t}{\partial \widehat{\sigma} \partial S}\, d\langle \widehat{\sigma}, S \rangle_t.
\end{aligned}$$
$$(3.115)$$

In the risk neutral world, the drift of the call must be equal to $rC_t dt$. Thus, by collecting the dt-terms in (3.115) and rearranging, the condition on the drift reads as

$$0 = \frac{\partial C_t}{\partial t} + (r-\delta)S_t \frac{\partial C_t}{\partial S} + \frac{1}{2}\widehat{\sigma}_t^2 S_t^2 \frac{\partial^2 C_t}{\partial S^2} - rC_t$$
$$+ \frac{1}{2}\left\{\sigma^2(S_t, t, \widehat{\sigma}_t) - \widehat{\sigma}_t^2\right\} S_t^2 \frac{\partial^2 C_t}{\partial S^2}$$
$$+ \alpha \frac{\partial C_t}{\partial \widehat{\sigma}} + \frac{1}{2}\frac{\partial^2 C_t}{\partial \widehat{\sigma} \partial \widehat{\sigma}}(\theta_0^2 + \theta_1^2) + \sigma(S_t, t, \widehat{\sigma}_t)\theta_0 S_t \frac{\partial C_t}{\partial S \partial \widehat{\sigma}}. \quad (3.116)$$

Obviously, the first line of (3.116) is the BS PDE (2.13) with IV replacing the volatility function. It must be equal to zero. Taking this into account, the condition on the drift is identified as

$$\alpha = \frac{1}{2}\left(\frac{\partial C_t}{\partial \widehat{\sigma}}\right)^{-1}\left[\left\{\widehat{\sigma}_t^2 - \sigma^2(S_t, t, \widehat{\sigma}_t)\right\} S_t^2 \frac{\partial^2 C_t}{\partial S_t^2}\right.$$
$$\left. - \frac{\partial^2 C_t}{\partial \widehat{\sigma} \partial \widehat{\sigma}}(\theta_0^2 + \theta_1^2) - 2\sigma(S_t, t, \widehat{\sigma}_t)\theta_0 S_t \frac{\partial C_t}{\partial S \partial \widehat{\sigma}}\right]. \quad (3.117)$$

Using the analytical derivatives of the BS call pricing formula given in (2.28) to (2.37), this reduces to

$$\alpha = \frac{1}{2\widehat{\sigma}_t \tau}\left\{\widehat{\sigma}_t^2 - \sigma^2(S_t, t, \widehat{\sigma}_t)\right\} - \frac{1}{2}\frac{d_1 d_2}{\widehat{\sigma}_t}(\theta_0^2 + \theta_1^2) + \frac{d_2}{\widehat{\sigma}_t \sqrt{\tau}}\sigma(S_t, t, \widehat{\sigma}_t)\theta_0, \quad (3.118)$$

which must be satisfied Q-almost surely to avoid arbitrage.

Equation (3.118) provides a number of interesting insights:

1. When IV is constant, i.e. $\theta_0 = \theta_1 = 0$, the instantaneous volatility σ must be a constant as well in order to satisfy (3.118).
2. If IV is a function only in time and strikes, i.e. again $\theta_0 = \theta_1 = 0$, the dynamics of the IV process reduce to $d\widehat{\sigma}_t = (2\tau)^{-1}\left\{\widehat{\sigma}_t^2 - \sigma^2(S_t, t, \widehat{\sigma}_t)\right\} dt$, which can be written as

$$\sigma^2(S_t, t, \widehat{\sigma}_t) = -\frac{d(\tau \widehat{\sigma}_t^2)}{dt}. \quad (3.119)$$

This in turn implies that the instantaneous volatility is as well only a function in time and strikes, as assumed in the (deterministic) local volatility models. Equation (3.119) relates back to the interpretation of the squared IV as the average squared volatility through the life time of the option. This was discussed in Sect. 2.8.
3. The drift is mean-fleeing as the first term shows on the right-hand side in (3.118). The further IV is away from instantaneous volatility, the further it is going to be pushed away. The speed of the mean-fleeing behavior increases as $T-t$ tends to zero, causing a 'volatility bubble', Schönbucher (1999).

3.12 Stochastic IV Models

The existence of the volatility bubble can be avoided by imposing restrictions on the instantaneous volatility as $T-t$ tends to zero. Indeed, it is easy to show that if the instantaneous volatility satisfies in the limit of $t \uparrow T$

$$-\widehat{\sigma}_t^4 + \widehat{\sigma}_t^2 \sigma^2(S_t, t, \widehat{\sigma}_t) - 2 x \theta_0 \widehat{\sigma}_t \, \sigma(S_t, t, \widehat{\sigma}_t) + x^2(\theta_0^2 + \theta_1^2) = 0 \, , \qquad (3.120)$$

where $x \stackrel{\text{def}}{=} -\ln \kappa_f = \ln\{e^{(r-\delta)\tau} S_t / K\}$ is (inverse) forward log-moneyness, bubbles are excluded from the model. This holds uniquely, since for $\theta_0, \theta_1, \sigma > 0$ and $x \in \mathbb{R}$ this polynomial has only one solution for $\widehat{\sigma}_t > 0$.

Equation (3.120) has at least two implications: first, it is seen that $\widehat{\sigma}_t$ is quadratic in x, which implies a *smile* across K. Since its shape is directly determined by θ_0 and θ_1, both parameters may be identified by calibration to the market smile. If $\theta_0 = 0$, i.e. if there is no correlation between the asset price and the IV dynamics, the smile is symmetric in x. Thus, asymmetry in the smile, the 'sneer', is introduced through the Brownian motion driving both variables. This parallels the work of Renault and Touzi (1996) as discussed in Sect. 2.11.

Second, the ATM IV defined in terms of the forward moneyness, i.e. where $x = 0$ converges to instantaneous volatility as $T - t$ tends to zero. However, this is *not* a consequence of the no-bubble restriction, but can be formally proved, Ledoit and Santa-Clara (1998); Daglish et al. (2003). This is seen as follows: from a first order Taylor series expansion of the BS pricing formula in the neighborhood of ATM (in the sense of log-forward moneyness), i.e. at $d_1 = -d_2 = \frac{1}{2}\widehat{\sigma}_t\sqrt{\tau}$, we obtain that

$$C_t(S_t, t, e^{(r-\delta)\tau} S_t, T) \approx \frac{1}{\sqrt{2\pi}} e^{-\delta\tau} S_t \widehat{\sigma}_t \sqrt{\tau} \, . \qquad (3.121)$$

This implies

$$\lim_{t \uparrow T} \widehat{\sigma}_t = \lim_{t \uparrow T} \sqrt{\frac{2\pi}{\tau} \frac{C_t}{e^{-\delta\tau} S_t}} \, . \qquad (3.122)$$

The call price can be approximated for small τ by

$$C_t = e^{-r\tau} \mathsf{E}^{\mathsf{Q}}\left\{(S_T - e^{(r-\delta)\tau} S_t)^+ | \mathcal{F}_t\right\}$$

$$\approx e^{-r\tau} \mathsf{E}^{\mathsf{Q}}\left\{S_t \sigma(S_t, t, \widehat{\sigma}_t)(\overline{W}_T^{(0)} - \overline{W}_t^{(0)})^+ | \mathcal{F}_t\right\}$$

$$= e^{-r\tau} S_t \sigma(S_t, t, \widehat{\sigma}_t) \sqrt{\frac{\tau}{2\pi}} \, , \qquad (3.123)$$

where the last line follows from the fact that $\mathsf{E}(z)^+ = \sqrt{\frac{\mathsf{Var}(z)}{2\pi}}$, where z is a normally distributed random variable with zero mean and variance $\mathsf{Var}(z)$. Inserting (3.122) and taking limits yields the desired result:

$$\lim_{t \uparrow T} \widehat{\sigma}_t = \lim_{t \uparrow T} \sigma(S_t, t, \widehat{\sigma}_t) \, . \qquad (3.124)$$

Note that this parallels the harmonic mean averaging result of Berestycki et al. (2002): here also, ATM local volatility, which is instantaneous volatility, converges to IV, Sect. 3.6.

The pricing of path-independent options works along standard lines. By standard results, the option price H must satisfy the following PDE subject to the appropriate boundary conditions:

$$0 = \frac{\partial H}{\partial t} + (r-\delta)S_t \frac{\partial H}{\partial S} + \frac{1}{2}\sigma^2(S_t,t,\widehat{\sigma}_t)S_t^2 \frac{\partial^2 H}{\partial S^2} - rH \\ + \sigma(S_t,t,\widehat{\sigma}_t)\theta_0 S_t \frac{\partial^2 H}{\partial \widehat{\sigma} \partial S} + \alpha \frac{\partial H}{\partial \widehat{\sigma}} + \frac{1}{2}(\theta_0^2 + \theta_1^2)\frac{\partial^2 H}{\partial \widehat{\sigma} \partial \widehat{\sigma}}. \tag{3.125}$$

Path-dependent options can be priced through Monte Carlo simulation.

In the implementation, difficulties may arise from the rather involved no-arbitrage conditions, Balland (2002). To simplify, Brace et al. (2001) propose to parameterize the volatility of IV as $\theta_i(\widehat{\sigma},t) \stackrel{\text{def}}{=} \theta_i \widehat{\sigma}$, $i=0,1$, where $\theta_i > 0$ is constant. This removes the singularities apparent in (3.118). Instead of obtaining the parameters from fitting the smile as suggested above, they can be recovered from PCA methods developed in Sect. 5.2. This path is taken for instance in Fengler et al. (2002b) and Cont et al. (2002). For the specification of the instantaneous volatility a lot of freedom remains, as long as (3.120) is satisfied in the limit. Alternatively, one may fix a drift function and recover from (3.118) the corresponding instantaneous volatility. For instance, the simplest choice would be to put $\alpha = 0$.

Finally, it should be remarked that the specification of the model in *absolute* terms, i.e. in terms of a fixed expiry date and a fixed strike may sometimes prove to be inconvenient in practice. Especially, an empirical identification of the parameters is likely to be more stable in terms of moneyness and time to maturity rather than in strikes and expiry dates. This is addressed in Brace et al. (2001) who show how to switch from the absolute to the *relative* notation of the model and its no-arbitrage conditions. Amerio et al. (2003) follow this approach and show how to price volatility derivatives using stochastic IV models.

3.13 Summary

In the first part of this chapter we introduced the theory of local volatility. Also several techniques for extracting local volatility from option prices were discussed. The focus was on implied trees. In the second part stochastic IV models were presented. At this point, we consider it to be appropriate to recall the concepts of volatility systematically. As explained in the introduction, we collected the main results in Fig. 3.11.

3.13 Summary 95

IV counterpart of Dupire formula (3.36)

local variance ——— determ. ——→ **implied variance**

$\sigma^2_{K,T}(S_t, t)$ no strike dependence $\widehat{\sigma}^2_t(K, T)$
or far OTM/ITM
arithmetic mean (2.78) and (3.47)

$t \uparrow T$: spatial harmonic mean of volatility (3.46)

$\mathsf{E}^{(K,T)}\{\sigma^2(S_T, T, \cdot)|\mathcal{F}_t\}$ $\{\mathsf{E}^{Q^{\lambda_1}}(\sqrt{\bar{\sigma}^2}|\mathcal{F}_t)\}^2$
Section 3.8 $K \approx F_t$, see (2.93)

$K = S_t, T = t$
see (3.4)

$K = F_t$, $t \uparrow T$
see (3.124)

instantaneous variance

$\sigma^2(S_t, t, \cdot)$

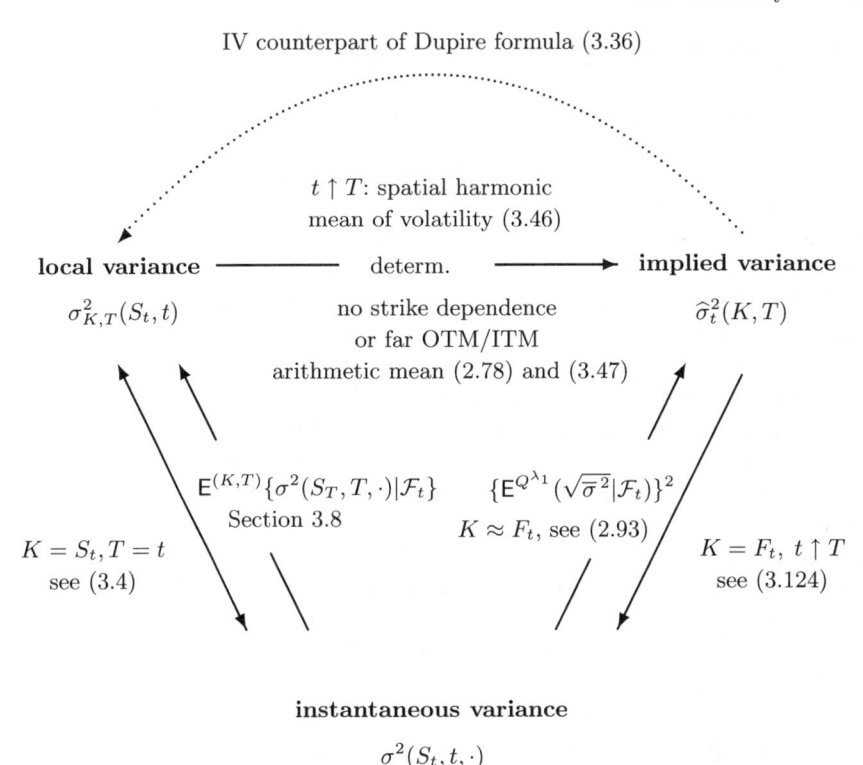

Fig. 3.11. Overview on volatility concepts. *Solid lines* denote exact relations between the different types of volatility. The *dashed line* denotes an ad-hoc concept. The arrows denote the direction of the relations

Starting from the most left arrow with the instantaneous variance, the first (and trivial) relation is the identity of local and instantaneous variance for $K = S_t$ and $T = t$. Moreover, local and implied variance can be represented as averages of instantaneous variance: local variance is the expectation under the (K, T)-risk adjusted measure. Implied variance is – for ATM options under the Hull and White (1987) model – the expectation under the risk neutral measure. Finally, the stochastic IV models show that ATM IV converges to instantaneous volatility as time to maturity converges to zero.

The asymptotic relations between implied and local volatility are presented in the top of the figure. They hold under the assumption of a deterministic instantaneous volatility function: first, IV is a spatial harmonic mean as time to maturity converges to zero. Second, if no strike dependence is present or for far OTM/ITM options, IV is a time average of local volatility. The Dupire formula in its IV representation – the dotted line – allows for recovering the LVS from the IVS and its derivatives. It is an ad-hoc concept, but a convenient way to reconstruct the LVS. Finally, the two-times-IV-slope was shown to hold for ATM options near to expiry.

After the recent theoretical and computational advances, local volatility models are found to be more and more criticized either for practical reasons or from theoretical grounds. From the practical perspective, there is the criticism that local volatility models deliver a wrong delta, Hagan et al. (2002). As discussed, the empirical literature does not appear to be strongly conclusive on the matter. A harsh methodological criticism is given by Ayache et al. (2004). Their main argument against local volatility is that these models lack economic grounds by not offering a reasonable smile explanation, as stochastic volatility or jump diffusion models do. Rather these models 'tweak' the diffusion coefficient in the BS PDE, until the observed option prices are matched. From their point of view, local volatility is just a computational construct bearing no economic content whatsoever. Given the highly spiky and counterintuitive surface structures that are typically recovered in local volatility models, this position cannot be completely dismissed. A somewhat milder position is taken by Wilmott (2001a, Chapter 25). He argues that local volatility models may be good when the options, from which the LVS is backed out, are simultaneously employed for static hedges: this can reduce the model error. In this case, one computes the LVS, and prices for instance a barrier option with respect to it. The option is then statically hedged by mimicking as close as possible the boundary condition and the payoff. However, we are not aware about a study simulating this strategy based on real data and assessing its success.

To summarize, given the current state of research, it seems to be difficult to give a concluding appraisal of local volatility models. The stochastic variants of local volatility or the stochastic IV models may offer fruitful solutions. But finally, it is daily practice on trading floors that has to determine whether local volatility can compete with stochastic volatility and jump-diffusions or not.

4
Smoothing Techniques

4.1 Introduction

Functional flexibility is a key requirement for model building and model selection in quantitative finance: often it is difficult, and sometimes impossible to justify on theoretical grounds a specific parametric form of an economic relationship under investigation. Furthermore, in a dynamic context, the economic structure may be liable to sizable changes and considerable fluctuations. Thus, estimation techniques that do not impose any a priori restrictions on the estimate, such as non- and semiparametric methods, are increasingly popular in financial practice.

In the case of the IVS, model flexibility is a prerequisite rather than an option: as has been seen in Chap. 2, from the BS theory, the IVS should be a flat and constant function across strike prices and the term structure of the option's time to maturity. Yet, as a matter of fact, one observes rich functional patterns fluctuating through time. This feature together with the discrete design, i.e. the fact that the daily IV observations occur only for a limited number of maturities, render IVS estimation an intricate challenge.

Parametric attempts to model the IVS along the strike profile, i.e. the 'smile', usually employ quadratic specifications, Shimko (1993), Ané and Geman (1999), and Tompkins (1999) among others. Also some of the methods listed for estimating the local volatility function are applicable here, Sect. 3.10.3. To allow for more flexibility, Hafner and Wallmeier (2001) fit quadratic splines to the smile function. However, it seems that these parametric approaches are not capable of capturing the salient features of IVS patterns, and hence estimates may be biased.

Recently, non- and semiparametric smoothing techniques for estimating the IVS have been used more and more: Aït-Sahalia and Lo (1998), Rosenberg (2000), Cont and da Fonseca (2002), Fengler et al. (2003b) employ a Nadaraya-Watson estimator of the IVS function, and higher order local polynomial smoothing of the IVS is used in Rookley (1997). Aït-Sahalia et al. (2001a) discuss model selection between fully parametric, semi- and nonparametric IVS specifications and argue in favor of the latter approaches.

The key idea in nonparametric estimation can be summarized as follows: suppose we are given a data set $\{(x_i, y_i)\}_{i=1}^n$, where $x_i \in \mathbb{R}$ denotes the predictor or the explanatory variables, and $y_i \in \mathbb{R}$ the response variable. In the context of IVS estimation, this would be some moneyness measure and time to maturity, or either of them, and IV respectively. The aim is to estimate the regression relationship

$$y_i = m(x_i) + \varepsilon_i, \quad i = 1, \ldots, n. \tag{4.1}$$

If one believes in some degree of smoothness between the explanatory variables and the response variable, it appears natural to assume that the data in the local neighborhood of a fixed point x contain information of m at x: thus, the basic idea in nonparametric estimation is to obtain an estimate $\widehat{m}(x)$ by *locally* averaging the data. More formally, this can be described by

$$\widehat{m}(x) = \frac{1}{n} \sum_{i=1}^{n} w_{i,n}(x) \, y_i, \tag{4.2}$$

where $\{w_{i,n}(x)\}_{i=1}^n$ denotes a sequence of weights. The weights reflect the likely fact that one will give higher weights to the observations x_i in the near vicinity of x than for those far off. Most nonparametric techniques can be written in this way, and differ only in the way the weights are computed.

In Sects. 4.2 and 4.3 of this chapter, we give an introduction into Nadaraya-Watson and local polynomial smoothing, which are the techniques employed for almost any of the graphical illustrations throughout this work. In Nadaraya-Watson smoothing one estimates a local constant, while in local polynomial smoothing one fits a polynomial of order p within a small neighborhood. From this point of view, Nadaraya-Watson smoothing is the special case of local polynomial smoothing with degree $p = 0$. Usually, in local polynomial smoothing, one uses a local linear estimator, i.e. $p = 1$, which is less affected by a bias in the boundary regions of the estimate than the Nadaraya-Watson estimator, Härdle et al. (2004). This however is asymptotically negligible.

When it is mandatory to also estimate derivatives, e.g. when the LVS is recovered from the IVS, Sect. 3.5, one needs to use higher order local polynomials. The degree of the polynomial depends on the number of derivatives desired. Since the local polynomial estimator can be written as a weighted least squares estimator, implementation is straightforward.

Section 4.5 presents an IVS estimator, a least squares kernel smoother, proposed by Gouriéroux et al. (1994) and Fengler and Wang (2003). This approach smoothes the IVS in the space of option prices and avoids a potentially undesirable feature of previous estimators: the *two*-step procedure. Traditionally, in a *first* step, IVs are derived by equating the BS formula with observed market prices and by solving for the diffusion coefficient, Sect. 2.5. In the *second* step the actual fitting algorithm is applied. A two-step estimator may be biased, when option prices or other input parameters can be observed with errors, only. Moreover, the nonlinear transformation of the option prices makes the error distribution less tractable. Indeed, it has been conjectured that the presence of measurement errors can be of substantial impact, see Roll (1984), Harvey and Whaley (1991), and particularly Hentschel (2003) for an extensive study on errors in IV estimation and their possible magnitude. Potential error sources are the bid-ask bounce, nonsynchronous pricing, infrequent trading of index stocks, and finite quote precision. Unlike the local polynomial smoother, the least squares kernel smoother does not have a closed-form solution, and for each grid point, the estimation must be achieved separately by a minimizing the objective function. On the other hand, as shall be seen, our results allow for the estimation of confidence bands that take the nonlinear transformation of the option prices into IVs into account.

A third methodology due to Fengler et al. (2003a) estimates the IVS via a semiparametric factor model. The reason for investigating this third approach lies in the very nature of the IVS data. As has been pointed out in Sect. 2.5, the IVS data are not equally distributed in the space, but occur in strings. Unless carefully calibrated, the fits obtained by the methods, which are discussed in this chapter, can be biased. The estimation strategy of the semiparametric factor model is specifically tailored to the degenerated, discrete string structure of the IVS data. It shall be discussed in Chap. 5, since we consider the dimension reduction aspects of this approach as its dominating feature, although it may also be seen as a pure estimation technique.

4.2 Nadaraya-Watson Smoothing

4.2.1 Kernel Functions

Nonparametric estimates are obtained by averaging the data locally. Usually, in this averaging, the data are given weights depending on the vicinity to $x \in \mathbb{R}$, at which the regression function m is to be found. The weighting is achieved by *kernel functions* $K(\cdot)$. The kernel functions employed in standard situations are continuous, positive, bounded and symmetric real functions which integrate to one:

$$\int K(u)\, du = 1 \, . \tag{4.3}$$

Kernel functions that are typically employed in nonparametric smoothing are the *quartic* kernel

$$K(u) = \frac{15}{16}(1-u^2)^2 \, \mathbf{1}(|u| \leq 1) \, , \tag{4.4}$$

and the *Epanechnikov* kernel

$$K(u) = \frac{3}{4}(1-u^2) \, \mathbf{1}(|u| \leq 1) \, , \tag{4.5}$$

which both have a bounded support. A kernel with infinite support is the *Gaussian* kernel, which is given by:

$$K(u) = \varphi(u) \stackrel{\text{def}}{=} \frac{1}{\sqrt{2\pi}} e^{-u^2/2}, \tag{4.6}$$

where $\pi = 3.141...$ denotes the circle constant.

For multidimensional smoothing tasks, as for IVS estimation, one needs multidimensional kernels. It is most common to obtain multidimensional kernels via products of univariate kernels:

$$K(u_1, \ldots, u_d) = \prod_{j=1}^{d} K_{(j)}(u_j), \tag{4.7}$$

which in this way inherit the properties of the univariate kernel function.

While different kernels have a different impact on the theoretical properties of the estimator, in practice the choice of the kernel function is not of big importance, and is mainly driven by practical considerations, Marron and Nolan (1988). For our work, we will only use quartic kernels and products of them.

The degree of localization or smoothing is steered via the bandwidth h. For instance, for a given data set $\{(x_i, y_i)\}_{i=1}^{n}$, for $x, y \in \mathbb{R}$, the bandwidths enter the kernel functions via

$$\frac{1}{h_n} K\left(\frac{x - x_i}{h_n}\right), \quad i = 1, \ldots, n. \tag{4.8}$$

The index n for the bandwidth clarifies that h_n actually depends on the number of observations. This is natural, since in the asymptotic perspective, as the number of observations tend to infinity, the degree of localization can shrink to zero without 'loosing' information about the regression function. In most cases, however, we will suppress this explicit notation.

Finally, it will occasionally be convenient to use the abbreviation

$$K_h(u) \stackrel{\text{def}}{=} \frac{1}{h} K\left(\frac{u}{h}\right). \tag{4.9}$$

4.2.2 The Nadaraya-Watson Estimator

For simplicity, consider the univariate model

$$Y = m(X) + \varepsilon, \tag{4.10}$$

with the unknown (but twice differentiable) regression function m. The explanatory variable X and the response variable Y take values in $\mathbb{R}$, have the joint pdf $f(x, y)$ and are independent of ε. The error ε has the properties $\mathsf{E}(\varepsilon|x) = 0$ and $\mathsf{E}(\varepsilon^2|x) = \sigma^2(x)$.

4.2 Nadaraya-Watson Smoothing

Taking the (conditional) expectation of (4.10) yields

$$\mathsf{E}(Y|X=x) = m(x) \,, \tag{4.11}$$

which says that the unknown regression function is the conditional expectation function of Y given $X = x$. Using the definition of the conditional expectation (4.11) can be written as

$$m(x) = \mathsf{E}(Y|X=x) = \frac{\int y f(x,y)\, dy}{f_x(x)} \,, \tag{4.12}$$

where f_x denotes the marginal pdf. Representation (4.12) shows that the regression function m can be estimated via the kernel density estimates of the joint and the marginal density. This approach was first introduced by Nadaraya (1964) and Watson (1964).

Suppose we are given the randomly sampled *iid* data set $\{(x_i, y_i)\}_{i=1}^n$. Then, the Nadaraya-Watson estimator is given by:

$$\widehat{m}(x) = \frac{n^{-1} \sum_{i=1}^n K_h(x - x_i)\, y_i}{n^{-1} \sum_{i=1}^n K_h(x - x_i)} \,. \tag{4.13}$$

Rewriting (4.13) as

$$\widehat{m}(x) = \frac{1}{n} \sum_{i=1}^n \frac{K_h(x - x_i)}{n^{-1} \sum_{j=1}^n K_h(x - x_j)} y_i = \frac{1}{n} \sum_{i=1}^n w_{i,n}(x)\, y_i \tag{4.14}$$

reveals that the Nadaraya-Watson estimator can be written as the locally weighted average of the response variable with weights

$$w_{i,n}(x) \stackrel{\text{def}}{=} \frac{K_h(x - x_i)}{n^{-1} \sum_{j=1}^n K_h(x - x_j)} \,. \tag{4.15}$$

Under some regularity conditions, the Nadaraya-Watson estimator is consistent, i.e.

$$\widehat{m}(x) \xrightarrow{p} m(x) \tag{4.16}$$

as $nh \uparrow \infty$, $h \downarrow 0$ and $n \uparrow \infty$. As opposed to parametric models under the correct specification, nonparametric estimates are biased. The bias, which is defined by $\text{Bias}\{\widehat{m}(x)\} = \mathsf{E}\{\widehat{m}(x) - m(x)\}$, can be reduced by decreasing the bandwidth, but this increases the variance of the regression function. The art in nonparametric regression lies trading off the variance and the bias.

The asymptotic bias of the Nadaraya-Watson estimator is

$$\text{Bias}\{\widehat{m}(x)\} = \frac{h^2}{2} \mu_2(K) \left\{ m''(x) + 2\frac{m'(x) f'_x(x)}{f_x(x)} \right\} \\ + \mathcal{O}(n^{-1} h^{-1}) + o(h^2) \,, \tag{4.17}$$

where $\mu_2(K) = \int u^2 K(u)\, du$, and the asymptotic variance is given by:

102   4 Smoothing Techniques

$$\text{Var}\{\widehat{m}(x)\} = \frac{1}{nh} \frac{\sigma^2(x)}{f_x(x)} \int K^2(u)\, du + o(n^{-1}h^{-1}) . \qquad (4.18)$$

For a precise treatment of the preceding statements see for instance Härdle (1990) or Pagan and Ullah (1999).

The Nadaraya-Watson estimator generalizes in a straightforward manner to the multivariate case: for some $\mathbb{R}^d$-valued sample $\{(\mathbf{x}_i, y_i)\}_{i=1}^n$, the multivariate Nadaraya-Watson estimator is given by

$$\widehat{m}(\mathbf{x}) = \frac{\sum_{i=1}^n K_\mathbf{h}(\mathbf{x}-\mathbf{x}_i)\, y_i}{\sum_{i=1}^n K_\mathbf{h}(\mathbf{x}-\mathbf{x}_i)} , \qquad (4.19)$$

where $K_\mathbf{h}(\cdot)$ denotes a multivariate kernel function with bandwidth vector $\mathbf{h} = (h_1, \ldots, h_d)^\top$. Similar results for the asymptotic bias and the asymptotic variance hold, Härdle et al. (2004).

4.3 Local Polynomial Smoothing

Another view on the Nadaraya-Watson estimator can be taken by noting that it can be written as the minimizer of (returning to the univariate case)

$$\widehat{m}(x) = \min_{m \in \mathbb{R}} \sum_{i=1}^n (y_i - m)^2 K_h(x-x_i) . \qquad (4.20)$$

Computing the normal equations of (4.20) leads to (4.13) as solution for m. This reveals that Nadaraya-Watson is a special case of fitting a constant in the local neighborhood of x. In local polynomial smoothing this idea is generalized to fitting locally a polynomial of order p.

Assume that the regression function is continuous up to order p. By expanding equation (4.10) in a Taylor series, we obtain

$$m(\xi) \approx m(x) + m'(x)(x-\xi) + \ldots + \frac{1}{p!} m^{(p)}(x)(x-\xi)^p \qquad (4.21)$$

for ξ in the neighborhood of x. Again we include the neighborhood of x via kernel weights. Thus, an estimator of $m(x)$ can be formulated in terms of the quadratic minimization problem

$$\min_{\boldsymbol{\beta} \in \mathbb{R}^{p+1}} \sum_{i=1}^n \left\{ y_i - \beta_0 - \beta_1(x-x_i) - \ldots - \beta_p(x-x_i)^p \right\}^2 K_h(x-x_i) , \qquad (4.22)$$

where $\boldsymbol{\beta} = (\beta_0, \ldots, \beta_p)^\top$ denotes the vector of coefficients. Obviously the result of this minimization problem is a weighted least squares estimator with weights $K_h(x_i - x)$.

We introduce the following matrix notation:

$$\mathbf{X} = \begin{pmatrix} 1 & x-x_1 & (x-x_1)^2 & \cdots & (x-x_1)^p \\ 1 & x-x_2 & (x-x_2)^2 & \cdots & (x-x_2)^p \\ \vdots & \vdots & \vdots & \ddots & \vdots \\ 1 & x-x_n & (x-x_n)^2 & \cdots & (x-x_n)^p \end{pmatrix}, \quad (4.23)$$

and $\mathbf{y} = (y_1, \ldots, y_n)^\top$, and finally

$$\mathbf{W} = \begin{pmatrix} K_h(x-x_1) & 0 & \cdots & 0 \\ 0 & K_h(x-x_2) & \cdots & 0 \\ \vdots & \vdots & \ddots & \vdots \\ 0 & 0 & \cdots & K_h(x-x_n) \end{pmatrix}. \quad (4.24)$$

Then we can write the solution of (4.22) in the usual least squares formulation as

$$\widehat{\boldsymbol{\beta}}(x) = (\mathbf{X}^\top \mathbf{W} \mathbf{X})^{-1} \mathbf{X}^\top \mathbf{W} \mathbf{y}. \quad (4.25)$$

Note that this estimator – unlike in the common parametric minimization schemes – varies in x, and therefore must be repeated for any x. This is highlighted by the notation $\widehat{\boldsymbol{\beta}}(x)$. The local polynomial estimator for the regression function is given by

$$\widehat{m}(x) = \widehat{\beta}_0(x), \quad (4.26)$$

by comparison of (4.21) and (4.22). From (4.25), writing the estimator as a local average of the response function is obvious.

Practice requires the choice of p. From the asymptotic behavior it is known that polynomials with odd degrees are to be preferred to those with even ones, i.e. the order one polynomial outperforms the order zero polynomial, the order three polynomial the order two polynomial etc. A case used particularly often is the local linear estimator with $p = 1$. It has been studied extensively by Fan (1992, 1993) and Fan and Gijbels (1992).

For the local linear estimator the asymptotic variance is identical to that stated in (4.18) for the Nadaraya-Watson estimator. The asymptotic bias takes the form:

$$\text{Bias}\{\widehat{m}(x)\} = \frac{h^2}{2} \mu_2(K) \, m''(x) + o(h^2). \quad (4.27)$$

Comparing (4.27) with (4.17), uncovers a remarkable difference: the bias does not depend on the densities, i.e. it is said to be *design adaptive*, Fan (1992). Moreover the bias vanishes, when m is linear. Thus local linear estimation can be superior to Nadaraya-Watson smoothing when the design becomes sparse as is typically the case for the IVS data. Another advantage of the local linear estimator is that its bias and variance are of the same order in magnitude in both the interior and the boundary of f_x. In practice, this may improve the behavior of the estimate near the boundary of the design.

An important byproduct of local polynomial estimators is that they provide an easy and efficient way for computing derivatives up to order $(p+1)$ of the regression function. For instance, the jth order derivative of m, $m^{(j)}$, is given by

$$\widehat{m}^{(j)}(x) = j!\, \widehat{\beta}_j(x) . \tag{4.28}$$

For the $\mathbb{R}^d$-variate extension to (4.22), one proceeds similarly. For instance, for the local linear estimator we have

$$\min_{\beta_0,\boldsymbol{\beta}_1 \in \mathbb{R}^d} \sum_{i=1}^n \left\{ y_i - \beta_0 - \boldsymbol{\beta}_1^\top (\mathbf{x} - \mathbf{x}_i) \right\}^2 K_\mathbf{h}(\mathbf{x} - \mathbf{x}_i) , \tag{4.29}$$

and a representation of the kind of (4.25) is obtained by introducing a suitable matrix notation, Härdle et al. (2004). For results on the asymptotic variance and bias see Ruppert and Wand (1994). Derivatives and cross-derivatives are obtained in the same way, i.e. by differentiating the local polynomial in $\boldsymbol{\beta}$ and by picking the appropriate vector entries.

4.4 Bandwidth Selection

4.4.1 Theoretical Framework

The crucial task in nonparametric smoothing is the bandwidth choice. This involves trading off the bias and the variance of the estimate. Typically, this is done by considering L^2-measures of distance between the estimate and the true regression curve. For the more detailed treatment of the following statements see Härdle (1990).

A way of balancing bias and variance in a pointwise sense is to minimize the mean squared error (MSE). The MSE is defined by:

$$\text{MSE}\{\widehat{m}(x)\} \stackrel{\text{def}}{=} \mathsf{E}[\{\widehat{m}(x) - m(x)\}^2] , \tag{4.30}$$

which can be written as

$$\text{MSE}\{\widehat{m}(x)\} = \text{Var}\{\widehat{m}(x)\} + [\text{Bias}\{\widehat{m}(x)\}]^2 , \tag{4.31}$$

where $\text{Bias}\{\widehat{m}(x)\} = \mathsf{E}\{\widehat{m}(x) - m(x)\}$.

Denote by AMSE the asymptotic MSE, which is obtained by ignoring all lower order terms in expressions like (4.17) and (4.18). This shows for the case of the Nadaraya-Watson estimates (as most other nonparametric estimates) that

$$\text{AMSE}(h) = \frac{1}{nh} c_1 + h^4 c_2 , \tag{4.32}$$

where c_1 and c_2 are constant. Minimizing with respect to h yields that

$$h \propto n^{-1/5} . \tag{4.33}$$

This, however, is of little use in practice, since the constants depend on unknown quantities like $\sigma^2(x)$ or $m''(x)$. Moreover, since the MSE is calculated for a specific point x only, it is a *local* measure. To reduce the dimensionality problem of optimizing h, one usually considers *global* measures.

A number of global measures can be defined. A typical choice is the integrated squared error (ISE):

$$\text{ISE}(h) \stackrel{\text{def}}{=} \int \{\widehat{m}(x) - m(x)\}^2 \widetilde{w}(x) f_x(x)\, dx\,, \tag{4.34}$$

where $\widetilde{w}(\cdot)$ is some weight function. It may be employed to assign less weight to regions where the data are sparse. A discrete approximation to the ISE is the average squared error (ASE):

$$\text{ASE}(h) \stackrel{\text{def}}{=} \frac{1}{n} \sum_{i=1}^{n} \{\widehat{m}(x_i) - m(x_i)\}^2 \widetilde{w}(x_i)\,. \tag{4.35}$$

Both the ISE and the ASE are random variables. Taking the expectation of the ISE, yields the mean integrated squared error (MISE)

$$\text{MISE}(h) \stackrel{\text{def}}{=} \mathsf{E}\{\text{ISE(h)}\}\,, \tag{4.36}$$

which is not a random variable. One may also take the expected value of the ASE, which yields the mean average squared error (MASE). We use a weighted version of the MASE for model selection in the semiparametric factor model, see Sect. 5.4.3.

For the Nadaraya-Watson estimator it has been shown by Marron and Härdle (1986, Theorem 3.4) that under mild conditions the ISE, ASE, and MISE are asymptotically equivalent in the sense that

$$\sup_{h} |\text{ASE}(h) - \text{MISE}(h)|/\text{MISE}(h) \longrightarrow 0 \quad \text{a.s.}\,, \tag{4.37}$$

and

$$\sup_{h} |\text{ISE}(h) - \text{MISE}(h)|/\text{MISE}(h) \longrightarrow 0 \quad \text{a.s.}\,, \tag{4.38}$$

for h from a closed set.

Still, we face the problem that these distance measures are not immediately computationally feasible in practice, since they depend on unknown quantities. However, the last (asymptotic) equivalence results open the way out by allowing to focus on the numerically most convenient of the three criteria – the ASE – and to suitably replace or to estimate the unknowns. This leads to *cross validation* and *penalizing techniques* as methods for the bandwidths choice. Both are based on a bias-corrected version of the *resubstitution estimate* of the prediction error:

$$p(h) \stackrel{\text{def}}{=} \frac{1}{n} \sum_{i=1}^{n} \{y_i - \widehat{m}(x_i)\}^2 \widetilde{w}(x_i)\,. \tag{4.39}$$

The cross validation function is defined by

$$CV(h) \stackrel{\text{def}}{=} \frac{1}{n}\sum_{i=1}^{n}\{y_i - \widehat{m}_{-j}(x_i)\}^2 \widetilde{w}(x_i) , \qquad (4.40)$$

where $\widehat{m}_{-j}$ denotes the *leave-one-out estimator* of the regression function, in which the jth observation is left out. For instance, for the case of the Nadaraya-Watson estimator, the leave-one-out estimator is given by

$$\widehat{m}_{-j}(x) = \frac{\sum_{i\neq j}^{n} K_h(x_j - x_i)\, y_i}{\sum_{i\neq j}^{n} K_h(x_j - x_i)} . \qquad (4.41)$$

In penalizing approaches one employs a weighted version of the resubstitution estimate:

$$G(h) = p(h) \, \Xi\left(\frac{1}{n} w_{i,n}(x_i)\right) \qquad (4.42)$$

with the correction function $\Xi(\cdot)$. It is required to have the first order Taylor expansion

$$\Xi(u) = 1 + 2u + \mathcal{O}(u^2), \quad u \to 0 . \qquad (4.43)$$

Typical choices of $\Xi(\cdot)$ are the *Akaike information criterion*:

$$\Xi_{AIC}(u) = \exp(2u) , \qquad (4.44)$$

and the *generalized cross validation selector*:

$$\Xi_{GCV}(u) = (1-u)^{-2} . \qquad (4.45)$$

For the generalized cross validation selector, we have $CV(h) = G(h)$ with Ξ_{GCV}. For other asymptotically equivalent choices of $\Xi(\cdot)$ see Härdle et al. (2004).

If we denote by $\widehat{h}$ the minimizer of $G(h)$ and by $\widehat{h}_*$ the minimizer of ASE, then for $n \uparrow \infty$

$$\frac{\text{ASE}(\widehat{h})}{\text{ASE}(\widehat{h}_*)} \stackrel{p}{\longrightarrow} 1 \quad \text{and} \quad \frac{\widehat{h}}{\widehat{h}_*} \stackrel{p}{\longrightarrow} 1 . \qquad (4.46)$$

Hence, independent of the specific choice of $\Xi(\cdot)$, the penalizing approach is asymptotically equivalent to the bandwidth obtained by minimizing the ASE.

4.4.2 Bandwidth Choice in Practice

Here, we give a short empirical demonstration of the estimators and explore the consequences of the bandwidth choice. For this application, we use option data from the dates 20010102 and 20010202. We explain in the Appendix A that call and put IVs can fall apart, when for the inversion of the BS formula, futures prices are used that are simply discounted. This is due to dividend effects and tax distortions, Hafner and Wallmeier (2001). It is best observed in late spring and early summer during the dividend season of the DAX index companies. To resolve this issue one applies a correction scheme, Appendix A.

In this section, we do not use the 'corrected data', since the least squares kernel estimator to be presented in the following employs a weight function that achieves this correction automatically by downweighing ITM options, which are most sensitive to the dividend wedge. Moreover, for the January and February data we use here this effect is hardly present. This implies that we can use the *simple* moneyness measure

$$\kappa \stackrel{\text{def}}{=} \frac{K}{S_t}, \qquad (4.47)$$

where $S_t = F_t e^{-r\tau}$, since $\delta \approx 0$.

In Table 4.1, we give an overview of the data employed. We prefer to present the summary statistics in form of the IV data obtained by inverting the BS formula separately for each observation rather than in form of the option price data itself. The corresponding option prices will be displayed later in the context of the least squares kernel estimator, see the top panel of Fig. 4.7. For the distribution of the data across moneyness compare Fig. 4.1, which presents density plots of moneyness for calls, puts, and all the observations observed on 20010102 for 17 days to expiry. The densities are obtained via a nonparametric density estimator, and bandwidths are chosen by Silverman's rule of thumb. Silverman's rule of thumb is a particular way to choose bandwidths in nonparametric density estimation, see Härdle et al. (2004) for details. Put and call densities appear shifted. This is due to the higher liquidity of ATM and OTM options. For the sake of space, we do not present the very similar plots for the other expiry dates and 20010202.

For our smile fits, we pick the options nearest to expiry from the 20010102 data. We start using the Nadaraya-Watson estimator for different bandwidths to demonstrate the tradeoff between bias and variance. The top left estimate

Table 4.1. IV data as obtained by inverting the BS formula separately for each observation in the sense of two-step estimators

Observation Date	Time to Expiry (Days)	Min	Max	Mean	Standard Deviation	Total Number of Observations	Calls
20010102	17	0.1711	0.3796	0.2450	0.0190	1219	561
	45	0.2112	0.2839	0.2425	0.0169	267	134
	73	0.1951	0.3190	0.2497	0.0199	391	209
	164	0.1777	0.3169	0.2528	0.0229	178	76
20010202	14	0.1199	0.4615	0.1730	0.0211	1560	813
	42	0.1604	0.2858	0.1855	0.0188	715	329
	77	0.1628	0.2208	0.1910	0.0172	128	45
	133	0.1645	0.2457	0.1954	0.0221	119	63

108   4 Smoothing Techniques

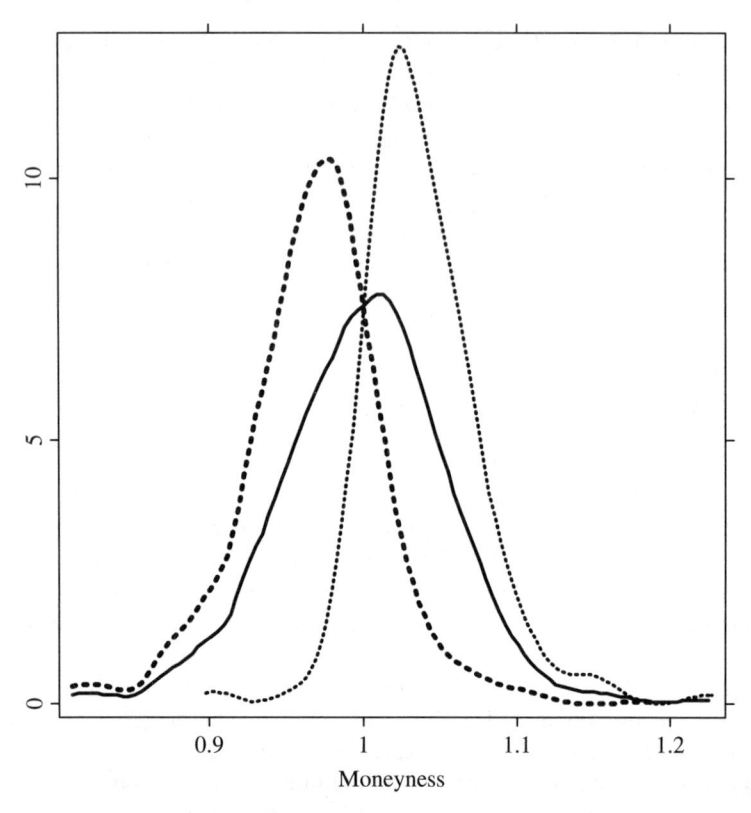

Fig. 4.1. Nonparametrically estimated densities of observed moneyness $\kappa = K/S_t$ for 20010102, and options with 17 days to expiry. Solid line for all observations, thickly dashed line for puts, and the more thinly dashed line for calls only. Quartic kernel used, bandwidth chosen according to Silverman's rule of thumb

for the bandwidth $h = 0.005$ in Fig. 4.2 is clearly undersmoothed: the estimate is very rough especially in the far OTM regions and has spikes. Since the smile in the ATM region looks already quite reasonable, one solution is to employ *local* bandwidths $h(x)$ that vary in x. In this case bandwidths should be an increasing function in either direction from ATM. Alternatively one may increase the global bandwidth: the estimate obtained for $h = 0.01$ appears already smoother, but still has some 'whiggles'. Increasing the bandwidth further to $h = 0.05$ yields the smooth smile function seen in the lower left panel in Fig. 4.2. However, the function appears already slightly biased, since in the wings of the smile the estimated function tends to lie systematically below the IV observations. This becomes more obvious for the large and extremely oversmoothing bandwidth $h = 0.1$, Fig. 4.2 lower right panel. The reason for this behavior of the Nadaraya-Watson estimator is that the number of observations become smaller and smaller the farther we move into the wings

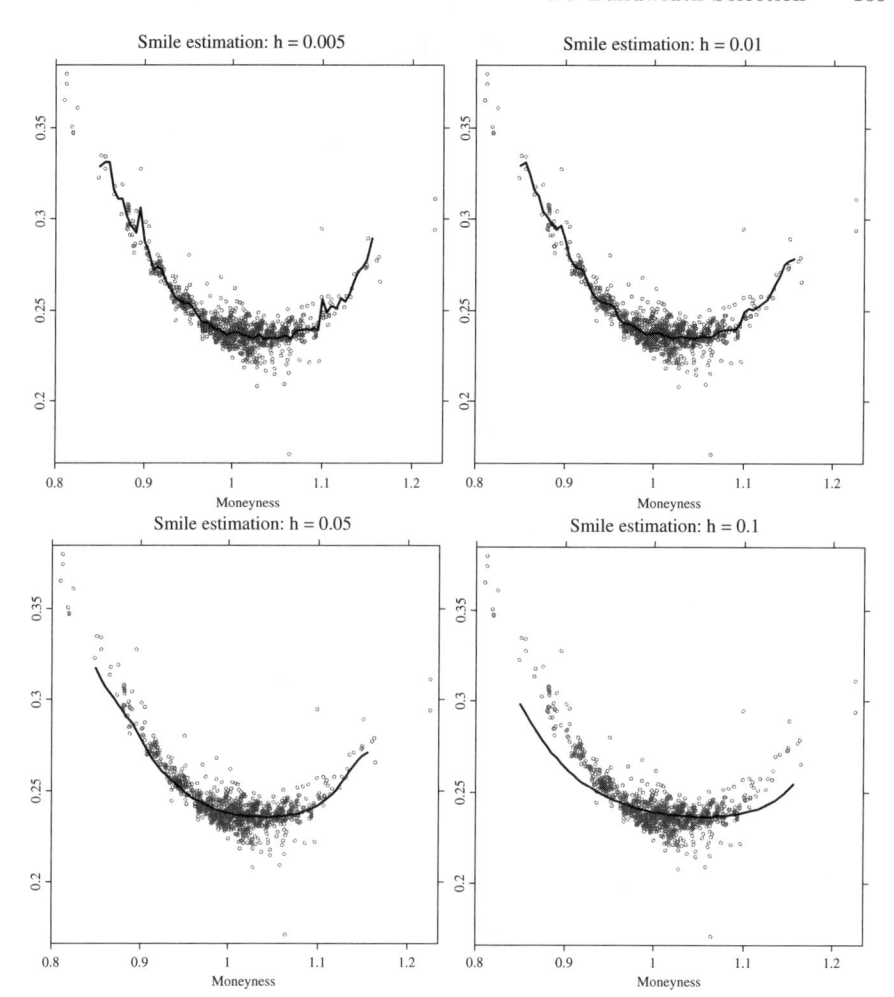

Fig. 4.2. Smile function obtained via Nadaraya-Watson smoothing for various bandwidths h. From *top left* to *lower right* h is: 0.005, 0.01, 0.05, 0.1

of the smile. Thus, within the local window of averaging, the estimate will be strongly influenced by the mass of the observations which have a lower IV.

Next, we run an Akaike penalizing approach for the bandwidth choice. In the top panel of Fig. 4.3, we display the penalized objective function. It is a convex function that takes its minimum in the neighborhood of 0.0285, for which we display the estimate in the lower panel. It appears to provide a reasonable fit to the data.

The exercise can be repeated for the local linear estimator. The results are displayed in Fig. 4.4. Typically the bandwidths for the local polynomial estimator need to be bigger than for the Nadaraya-Watson estimator. This is seen in the upper left plot of the figure. Here, the bandwidth in the wings

110   4 Smoothing Techniques

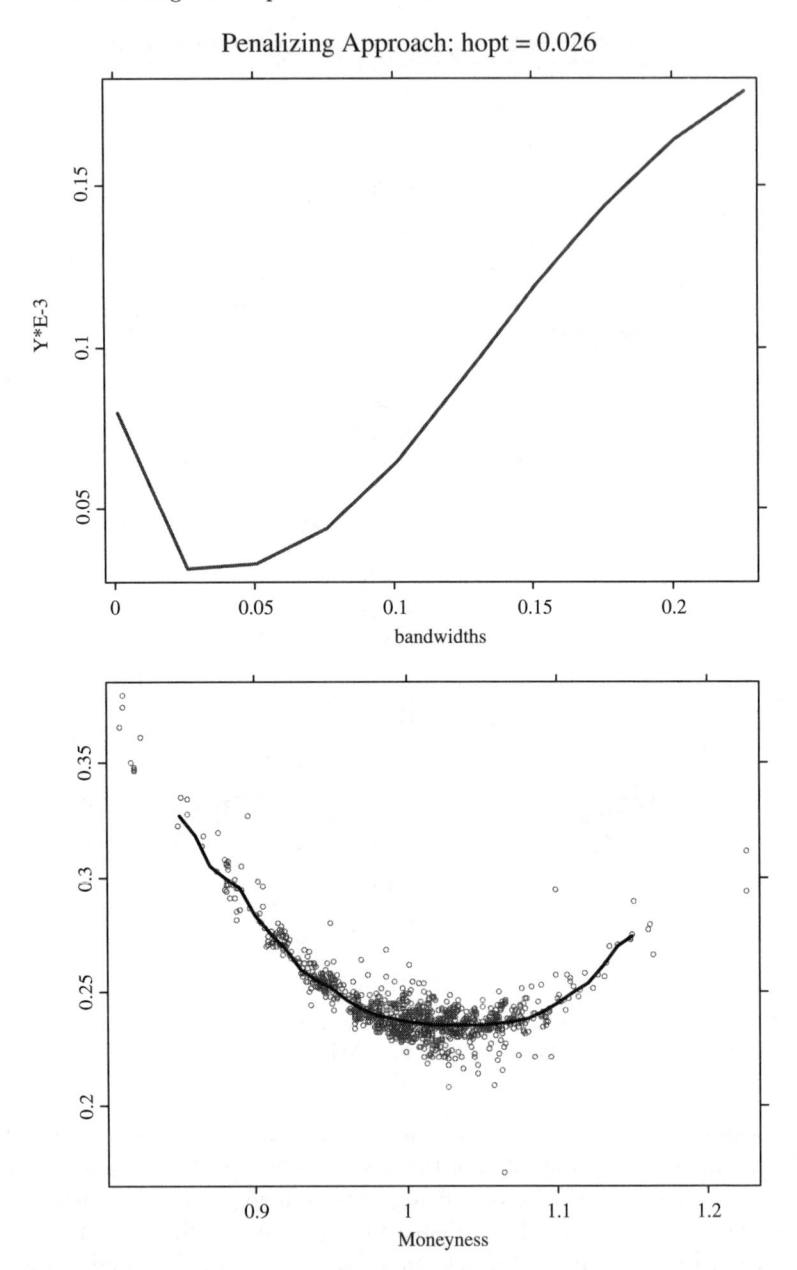

Fig. 4.3. The *top panel* displays the penalized resubstitution estimate of the Nadaraya-Watson estimator. Penalizing function is the Akaike function (4.44). The *lower panel* shows the smile function obtained for the optimal bandwidth $h = 0.028$

4.4 Bandwidth Selection

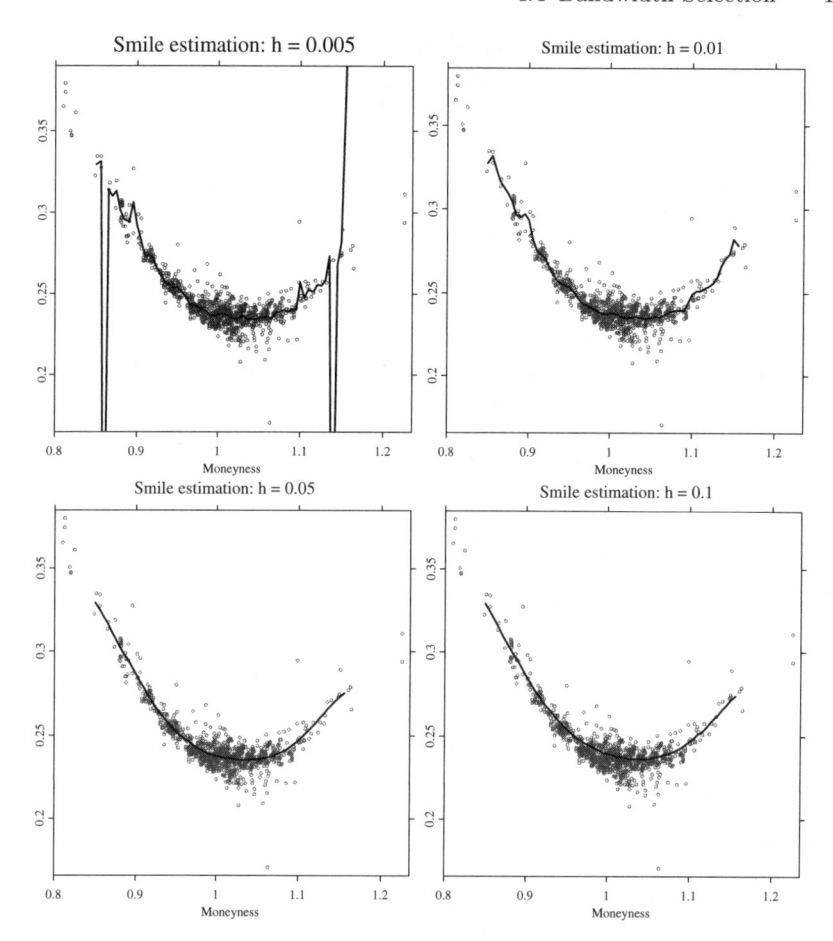

Fig. 4.4. Smile function obtained via local linear smoothing for various bandwidths h. From *top left* to *lower right* h is: 0.005, 0.01, 0.05, 0.1

of the smile is too small to yield a reasonable estimate. For the bigger bandwidths better estimates are obtained. Note that the bias problem visible for the Nadaraya-Watson estimator is less present in local linear smoothing: even for the biggest bandwidth from our set, 0.1, we receive a reasonable result. This is because even for larger intervals, the IV smile can be reasonably well fitted by piecewise linear splines. The bandwidth needs to be increased much stronger to produce an estimate similar to that in the lower right panel of Fig. 4.2. Given the typical parabolic shape of the smile function, this effect is even more striking for local quadratic fits.

For precisely this reason we prefer *local polynomial smoothing* in smile modeling: for the functional shapes that are usually encountered in smile modeling, the local polynomial estimates appear to be relatively *robust* against oversmoothing. This facilitates bandwidth choice enormously for two reasons:

first, the data in the outer wings of the smile can become very sparse. Thus, if a global bandwidth is to be used, it is likely that the smile needs to be oversmoothed. Second, from the perspective of computing daily estimates of the smile in a large sample as ours, it can be justified to employ one single and potentially slightly oversmoothing bandwidth for all estimates without minimizing the penalized resubstitution estimate again and again.

For estimates of the entire IVS, in principle one could proceed similarly. The empirical difficulty, however, seems to be that both cross validation and penalizing approaches tend to yield unsatisfactory results due to the intricate design of the IV data in the time to maturity direction: while the bandwidth optimization in moneyness direction poses no difficulty, adding the time to maturity dimension leads to convexity problems in the penalized function and consequently to unreasonable minimizers, such as boundary solutions. This phenomenon has been first discussed by Fengler et al. (2003b), see also Fengler et al. (2003a).

The practical solution we adopt in most cases, where we use global bandwidths, such as in the CPC analysis, is the following: we run the aforementioned minimization only across moneyness in each of a number of daily samples. Next, we inspect the minimizers and the bias over a wide range of bandwidths. Typically the conclusions are similar, and we use slightly oversmoothing, but fixed bandwidths for all estimates. This approach is justified by the fact that in the time to maturity direction one is more interested in *interpolation* rather than in *smoothing*. In the semiparametric factor model, where visible inspection is not directly possible, we propose a weighted Akaike penalization that explicitly takes into account the sparseness of the data. This is explained in Sect. 5.4.

In Fig. 4.5, we present two surfaces, a Nadaraya-Watson and a local linear estimate, for the data of 20010102. It is again visible that the local linear estimator captures better the smile form, especially for IVs near to expiry.

Before concluding let us revisit the estimate of the LVS in Sect. 3.1 recovered from the IV counterpart of the Dupire formula, Equation (3.39). This requires estimating the first order derivatives of the IVS with respect to moneyness and time to maturity as well as the second order derivative with respect to moneyness. This can be achieved by local polynomial smoothing of order $p \geq 2$. Here, we employ an order two polynomial. Moreover, in order to achieve an exact fit of the data we use *local* bandwidths $h(x)$. This can be achieved by employing a smooth function $h(x)$ that approximates the global bandwidths that have been obtained from separate cross validations for the short and long time to maturity data. A more sophisticated method is an empirical bias-bandwidth selection procedure, Ruppert (1997). Local bandwidths allow for better capturing the gap between the short and the long term time to maturity strings. In Fig. 4.6, we present the derivatives together with the regression function itself, from which – via the moneyness representation of the Dupire formula (3.39) – the LVS is recovered.

4.4 Bandwidth Selection   113

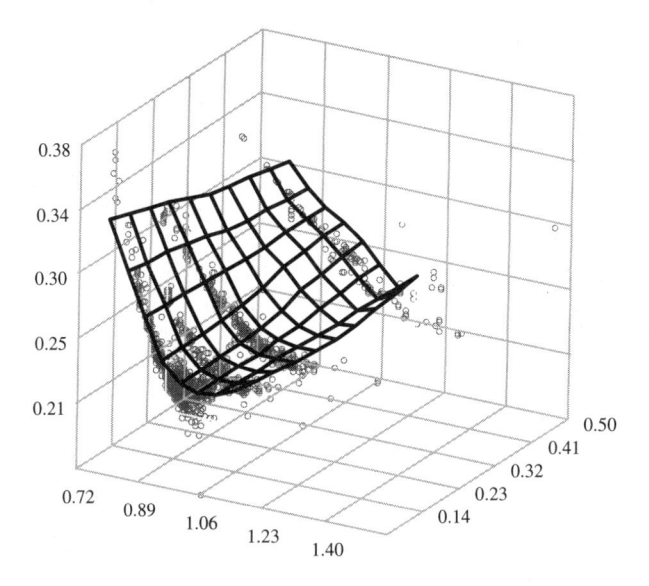

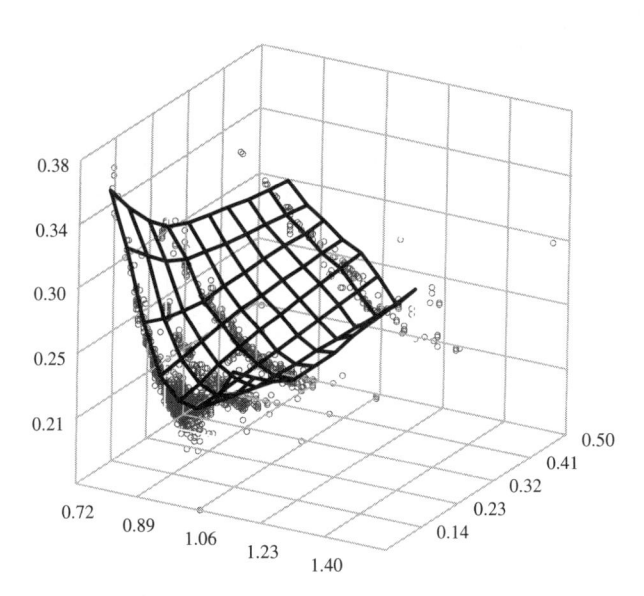

Fig. 4.5. IVS estimation via Nadaraya-Watson (*top panel*) and local linear smoothing (*lower panel*). Global bandwidth $h_1 = 0.04$ in moneyness and $h_2 = 0.3$ in time to maturity direction

114   4 Smoothing Techniques

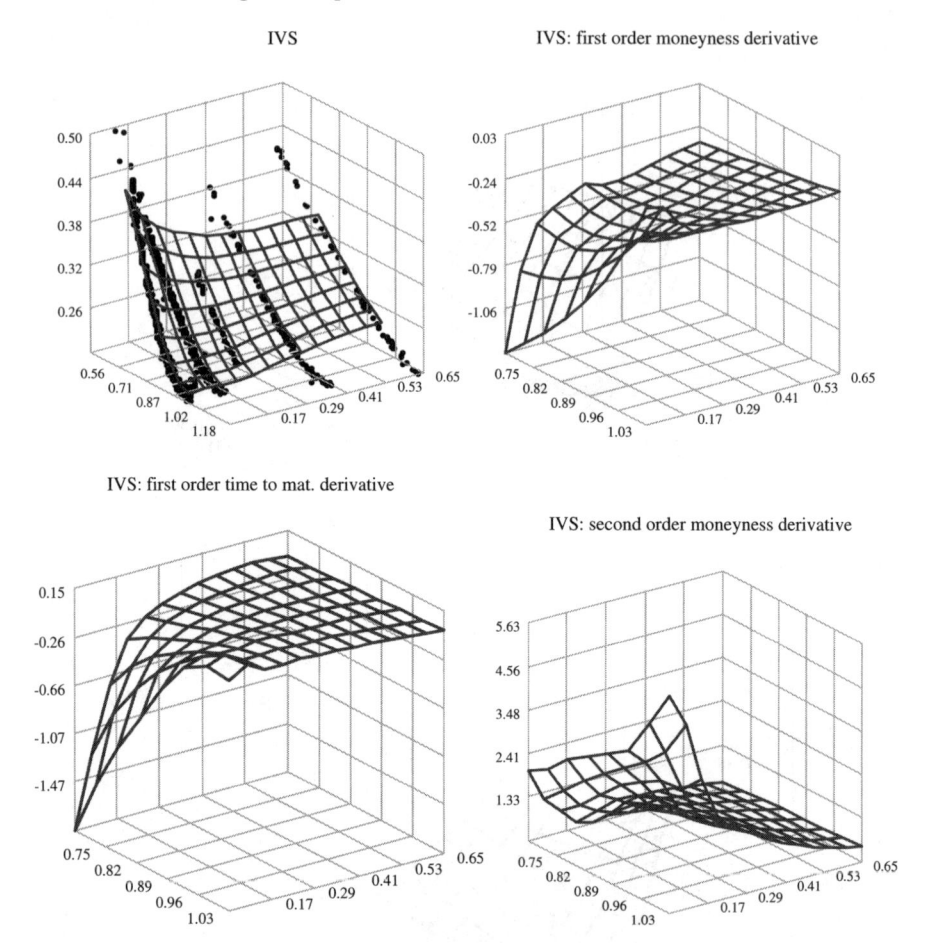

Fig. 4.6. IVS derivative estimation via local polynomial estimation of order two. From *upper left* to *lower right*, the plots show the IVS on 20000502, the first order moneyness derivative, the first order time to maturity derivative, and the second order moneyness derivative. Bandwidths are localized. The corresponding LVS plot was given in Fig. 3.1

4.5 Least Squares Kernel Smoothing

4.5.1 The LSK Estimator of the IVS

In this section, we propose a special smoother designed for estimating the IVS. It is a one-step procedure based on a least squares kernel (LSK) estimator that smoothes IV in the space of option prices. There is no need for first inverting the BS formula to recover IV observations – the observed option prices are the input parameters required. The LSK estimator is a special case of a general class of estimators, the so called kernel M-estimators, that has been introduced by Gouriéroux et al. (1994). Gouriéroux et al. (1995) employ this estimator to model and predict stochastic IV.

Since we aim at estimating on a moneyness metric, we rewrite the BS formula for calls (2.23) in terms of moneyness as follows, Gouriéroux et al. (1995):

$$C^{BS}(S_t, t, K, T, \sigma, r, \delta) = S_t c^{BS}(\kappa_t, \tau, \sigma, r, \delta) , \qquad (4.48)$$

where $c^{BS}(\kappa_t, \tau, \sigma, r, \delta) \stackrel{\text{def}}{=} \Phi(d_1) - \kappa_t e^{-r\tau} \Phi(d_2)$, and $d_1 = \frac{-\ln \kappa_t + (r + \frac{1}{2}\sigma^2)\tau}{\sigma\sqrt{\tau}}$, $d_2 = d_1 - \sigma\sqrt{\tau}$ as before. We recall that throughout this section we work with the simple moneyness measure

$$\kappa_t \stackrel{\text{def}}{=} \frac{K}{S_t} . \qquad (4.49)$$

We add a subscript t in order to highlight the time dependence. For simplicity, we shall also assume zero dividends. Inserting a constant dividend yield would be a simple extension.

The LSK estimator for the IVS is defined by:

$$\widehat{\sigma}(\kappa_t, \tau) = \arg\min_{\sigma} \sum_{i=1}^{n} \left\{ \widetilde{c}_{t_i} - c^{BS}(\cdot, \sigma) \right\}^2$$
$$\times w(\kappa_{t_i}) K_{(1)}\left(\frac{\kappa_t - \kappa_{t_i}}{h_1}\right) K_{(2)}\left(\frac{\tau - \tau_i}{h_2}\right) . (4.50)$$

Here, the observed call prices are normalized by the asset price, i.e. $\widetilde{c}_t \stackrel{\text{def}}{=} \widetilde{C}_t/S_t$, for $i = 1, \ldots, n$. Otherwise notation stays as introduced in Sect. 2.4. $K_{(1)}(\cdot)$ and $K_{(2)}(\cdot)$ are univariate kernel functions, and $w(\cdot)$ denotes a uniformly continuous and bounded weight function, which allows for differential weights of observed option prices. This weight function is useful in the following respect: it is usually argued that ITM options contain a liquidity premium and should be incorporated to a lesser extent into the IV estimate, or even excluded, Aït-Sahalia and Lo (1998) and Skiadopoulos et al. (1999). This goal can be achieved in using an appropriate weight function $w(\kappa)$. In Sect. 4.5.2, we will discuss plausible choices of the weight functions.

We make the following assumptions:

(A1) The moneyness of the option prices is *iid*, and $\mathsf{E}\kappa_t^4 < \infty$.
(A2) The weight function $w(\cdot)$ is uniformly continuous and bounded.
(A3) $K_{(1)}(\cdot)$ and $K_{(2)}(\cdot)$ are bounded probability density kernel functions with bounded support.
(A4) Interest rate r is a fixed constant.

Assumption (A1) is a very weak assumption. It can very well be considered to hold in practice, since by the institutional arrangements at futures exchanges, options at new strikes are always launched in the neighborhood of S_t. The proof relies on $\mathsf{E}(\widetilde{c}_t^4|\mathcal{F}_t) = \mathsf{E}\{(\widetilde{C}_t/S_t)^4|\mathcal{F}_t\} < \infty$. However, since S_t is measureable with respect to $\mathcal{F}_t$ and by simple no-arbitrage considerations, we have $0 \leq \mathsf{E}_t(\widetilde{C}_t|\mathcal{F}_t) \leq S_t$. So this condition is implied by option pricing theory.

Assumption (A2) is very common, and some important weight function satisfy it. In Sect. 4.5.2 we will discuss possible choices of $w(\cdot)$. (A3) is a condition met by a lot of kernels used in nonparametric regression, such as the quartic or the Epanechnikov kernel functions, Sect. 4.2.1. (A4) is an assumption often used in the option pricing literature including the BS model. It is generally justified by the empirical observation that asset pricing variability largely outweighs the changes of the interest rates. Nevertheless, the impact from changing interest rates can be substantial for options with a very long time to maturity.

Given assumptions (A1) to (A4), we have:

Consistency. *Let $\sigma(\kappa_t, \tau)$ be the solution of*
$\mathsf{E}[\{\widetilde{c}_{t_1} - c^{BS}(\kappa_t, \tau, r_1, \sigma)\}w(\kappa_t)|\mathcal{F}_t] = 0$. *If conditions (A1), (A2), (A3) and (A4) are satisfied, then*

$$\widehat{\sigma}(\kappa_t, \tau) \xrightarrow{p} \sigma(\kappa_t, \tau)$$

as $nh_{1,n}h_{2,n} \to \infty$.

The proof can be found in Gouriéroux et al. (1994) and is contained for the sake of completeness in the version of Fengler and Wang (2003) in Appendix C.

For the next result, we introduce the notation:

$$A_i(\kappa_t, \tau, r, \sigma) \stackrel{\text{def}}{=} \widetilde{c}_{t_i} - c^{BS}(\kappa_{t_i}, \tau_i, r_i, \sigma) ,$$

$$B(\kappa_t, \tau, r, \sigma) \stackrel{\text{def}}{=} \frac{\partial c^{BS}(\cdot)}{\partial \sigma} = S_t^{-1}\frac{\partial C^{BS}(\cdot)}{\partial \sigma} = \sqrt{\tau}\phi(d_1) , \qquad (4.51)$$

$$D(\kappa_t, \tau, r, \sigma) \stackrel{\text{def}}{=} \frac{\partial^2 c^{BS}(\cdot)}{\partial^2 \sigma} = S_t^{-1}\frac{\partial^2 C^{BS}(\cdot)}{\partial \sigma^2} = \sqrt{\tau}\phi(d_1)\frac{d_1 d_2}{\sigma} , \qquad (4.52)$$

4.5 Least Squares Kernel Smoothing

The quantities B and D are the 'moneyness versions' of the vega and the volga, which have been introduced in Sect. 2.4.

Asymptotic normality. *Under conditions (A1), (A2), (A3), and (A4) if $E\{B^2(\kappa_t,\tau,r,\sigma)w(\kappa_t)|\mathcal{F}_t\} \neq E\{A(\kappa_t,\tau,r,\sigma)D(\kappa_t,\tau,r,\sigma)w(\kappa_t)|\mathcal{F}_t\}$, we have*

$$\sqrt{nh_{1,n}h_{2,n}}\{\widehat{\sigma}(\kappa_t,\tau) - \sigma(\kappa_t,\tau)\} \xrightarrow{\mathcal{L}} N(0,\gamma^{-2}\nu^2),$$

where

$$\gamma^2 \stackrel{\text{def}}{=} \Big[E\{-B^2(\kappa_t,\tau,r,\sigma)w(\kappa_t) \\ + A(\kappa_t,\tau,r,\sigma)D(\kappa_t,\tau,r,\sigma)w(\kappa_t)|\mathcal{F}_t\}\Big]^2 f_t(\kappa_t,\tau) , \quad (4.53)$$

$$\nu^2 \stackrel{\text{def}}{=} E\{A^2(\kappa_t,\tau,r,\sigma)B^2(\kappa_t,\tau,r,\sigma)w^2(\kappa_t)|\mathcal{F}_t\} \\ \times \int K_{(1)}^2(u)K_{(2)}^2(v)\,dudv , \quad (4.54)$$

and $f_t(\kappa_t,\tau)$ is the joint (time-t conditional) probability density function of κ_t and τ respectively.

For the proof see Gouriéroux et al. (1994) and in Appendix C. Finally, the results carry over to put options: By the put-call-parity and the bounded pay-off of put options, both results hold also for put options, with A replaced correspondingly.

The asymptotic distribution depends intricately on the first and second order derivatives, and the particular weight function. Nevertheless an approximation is simple, since the first and second order derivatives have the analytical expressions given in Equations (4.51) and (4.52).

4.5.2 Application of the LSK Estimator

For the choice of the weighting function, one may go back to the early literature on IV. In the vain of obtaining a good forecast of the asset price variability, these studies discuss weighting the observations intensively, see the discussion in Sect. 2.10. Schmalensee and Trippi (1978) and Whaley (1982) argue in favor of unweighted averages, i.e. they use the scalar estimate

$$\widehat{\sigma} = \arg\min_{\sigma} \sum_{i=1}^{n}\{\widetilde{C}_i - C^{BS}(\cdot,\sigma)\}^2 , \quad (4.55)$$

as a predictor of the future stock price variability. Beckers (1981) minimizes

$$\widehat{\sigma} = \arg\min_{\sigma} \sum_{i=1}^{n} w_i \{\widetilde{C}_i - C^{BS}(\cdot, \sigma)\}^2 / \sum_{i=1}^{n} w_i \,, \qquad (4.56)$$

where $w_i \stackrel{\text{def}}{=} \partial C_i / \partial \sigma$ is the option vega.

Similarly, Latané and Rendelman (1976) use the squared vega as weights:

$$\widehat{\sigma} = \sqrt{\sum_{i=1}^{n} w_i^2 \widehat{\sigma}_i^2 / \sum_{i=1}^{n} w_i} \,. \qquad (4.57)$$

Finally, Chiras and Manaster (1978) propose to employ the elasticity with respect to volatility:

$$\widehat{\sigma}^* = \sum_{i=1}^{n} \eta_i \widehat{\sigma}_i / \sum_{i=1}^{n} \eta_i \,, \qquad (4.58)$$

where $\eta_i \stackrel{\text{def}}{=} \frac{\partial C_i}{\partial \sigma} \frac{\sigma}{C_i}$.

For calls and puts, vega is a Gaussian shaped function in the underlying centered (roughly) ATM, compare Equation (4.51) and Fig. 2.3. Elasticity is a decreasing (increasing) function in the underlying for calls (puts). Common concern of the weighting procedures is to give low weight to ITM options, and highest weight to ATM or OTM options: ITM options are more expensive than ATM and OTM options because their intrinsic value, i.e. the payoff function evaluated at the current underlying prices, is already positive. Thus, they provide lower leverage for speculation, and produce higher costs in portfolio hedging. Due to their lower trading volume, they are suspected to sell at a liquidity premium from which biased estimates of IV may ensue. Consequently, some authors delete or downweigh ITM options, Aït-Sahalia and Lo (1998).

The LSK estimator is general enough to allow for uniformly continuous and bounded weighting functions $w(\kappa)$ depending on moneyness. Technically, it is possible to use weights depending also on other variables including σ as done in (4.56) to (4.58). For several reasons, however, we refrain from using more involved weight functions: first, when ITM options are deleted or downweighted in the more recent literature, this choice is entirely determined by moneyness, not by the vega. From this point of view, to have the weighting scheme depend on σ is rather implicit. Second, from a statistical point of view, weights depending on σ are likely to blow up the asymptotic variances in form of the derivatives of w. This complicates the estimation and the computation of the confidence bands without adding to the problem of recovering a good estimate of the IVS. Finally, if one likes weights looking like the option vega or elasticity with respect to volatility, one may very easily construct weights $w(\kappa)$ that look very similar. For instance, an estimator in the type of Latané and Rendelman (1976) would put w shaped as a Gaussian density.

4.5 Least Squares Kernel Smoothing

For the IVS estimation in our particular application, we want to give less weight to ITM options. This can be achieved by using as weighting functions:

$$w(\kappa) = \frac{1}{\pi}\arctan\{\alpha(1-\kappa)\} + 0.5, \qquad (4.59)$$

for calls, and for puts:

$$w(\kappa) = \frac{1}{\pi}\arctan\{\alpha(\kappa-1)\} + 0.5, \qquad (4.60)$$

where $\pi = 3.141...$ is the circle constant. The parameter α controls the speed, with which ITM options receive lower weight. ATM options are equally weighted. Outside $\kappa \approx 1$, only OTM options enter the minimization with significant weight. In our application we choose $\alpha = 9$. Other values are perfectly possible, and this choice is motivated to have a gentle transition between OTM call and OTM put options. The ultimate choice of α will depend on the specific application at hand.

As kernel functions we employ the quartic kernels given in (4.4). Other bounded kernels can perfectly be used, such as the Epanechnikov kernel as stated in (4.5). In practice, the choice of the kernel functions has little impact on the estimates, Marron and Nolan (1988) and Härdle (1990). Since the minimization is globally convex (compare the proof of consistency in the appendix), and well posed as long as h_1 and h_2 do not become unreasonably small, any minimization algorithm for globally convex objective functions can be employed. We use the Golden section search, described for instance in Press et al. (1993) and implemented in XploRe, Härdle et al. (2000b). The tolerance, i.e. the fractional precision of the minimum, is fixed at 10^{-8}.

We use the same data as already presented in Table 4.1. For the smile estimation, we pick the options with the shortest time to expiry from the 20010102 and the 20010202 data. Plots are displayed in Figs. 4.7 and 4.8. The top panel shows the observed option prices given on the moneyness scale. The function from the lower left to the upper right is the put price function, the one from the upper left to the lower right the call price function. This is at odds with the familiar ways of plotting these functions. The effect is due to our definition of moneyness. The lower panels in Figs. 4.7 and 4.8 present the smile together with the asymptotic confidence bands. They fan out at the wings of the smile since the data become increasingly sparse.

In Fig. 4.9, fits for the entire IVS are presented. They appear undersmoothed compared with Fig. 4.5, since we used very small bandwidths in the time to maturity direction. For these estimates, we do not employ the weight functions (4.59) and (4.60): all observations are equally weighted.

120   4 Smoothing Techniques

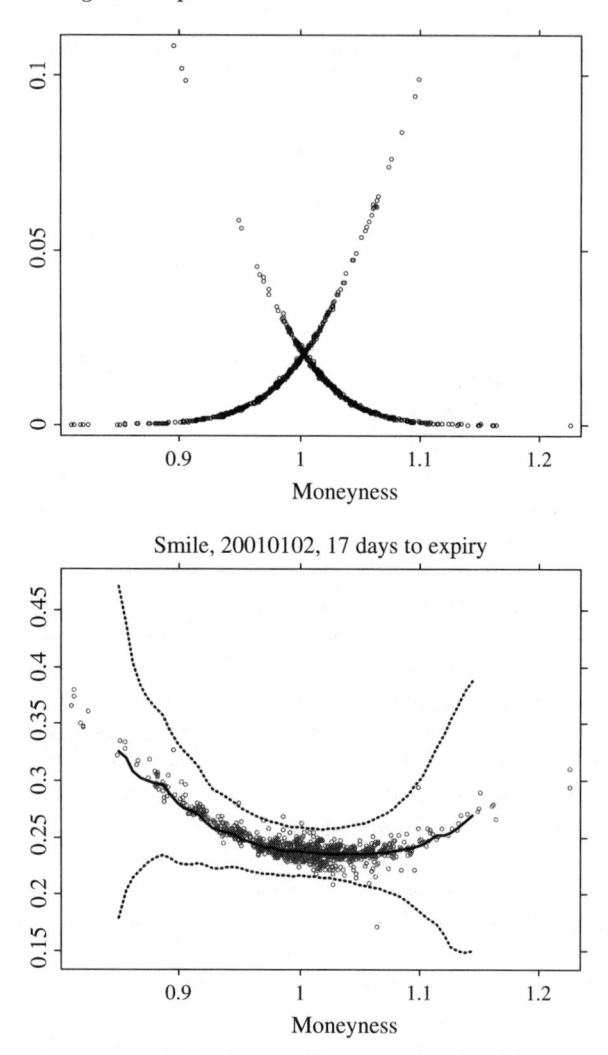

Fig. 4.7. *Upper panel*: Observed option price data on 20010102. From *lower left* to *upper right* relative put prices, from *upper left* to *lower right* relative call prices. *Lower panel*: LSK smoothed IV smile for 17 days to expiry on 20010102. Bandwidth $h_1 = 0.025$, quartic kernels employed. Minimization achieved by Golden section search. *Dotted lines* are the 95% confidence intervals for $\hat{\sigma}$. Single dots are IV data obtained by inverting the BS formula separately for each observation in the sense of two-step estimators

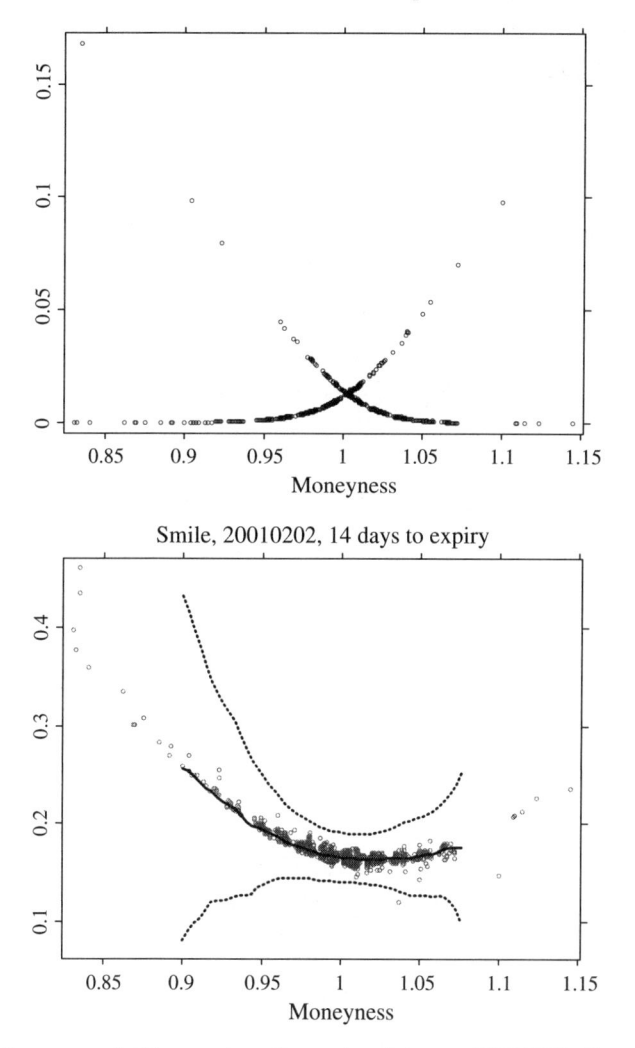

Fig. 4.8. *Upper panel*: Observed option price data on 20010202. From *lower left* to *upper right* relative put prices, from *upper left* to *lower right* relative call prices. *Lower panel*: LSK smoothed IV smile for 14 days to expiry on 20010202. Bandwidth $h_1 = 0.015$, quartic kernels employed. Minimization achieved by Golden section search. *Dotted lines* are the 95% confidence intervals for $\hat{\sigma}$. Single dots are IV data obtained by inverting the BS formula separately for each observation in the sense of two-step estimators

122   4 Smoothing Techniques

IVS on Jan. 02, 2001

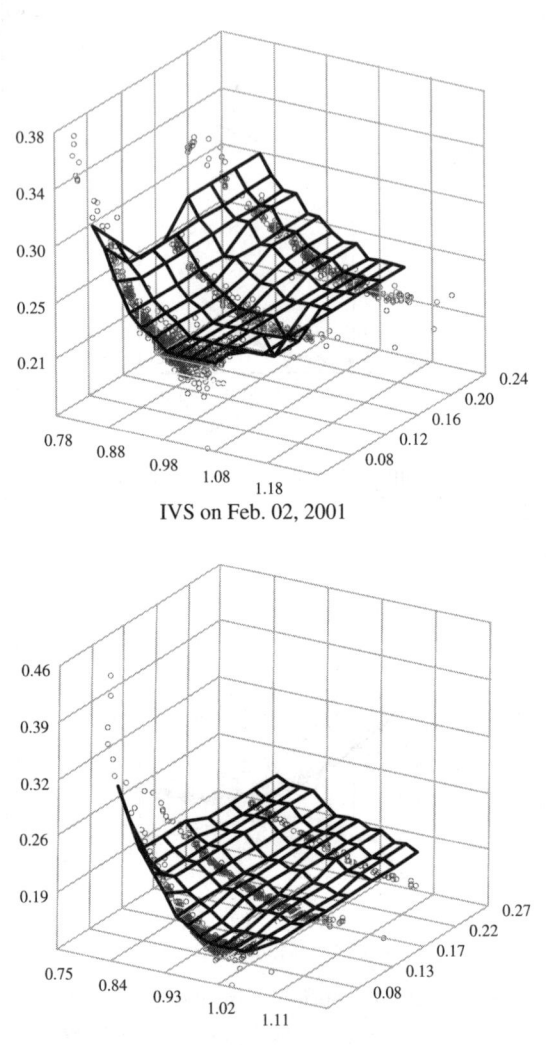

IVS on Feb. 02, 2001

Fig. 4.9. *Top panel*: IVS fit for 20010102; *lower panel*: IVS fit for 20010202, both with the LSK smoother. In both panels, bandwidths are $h_1 = 0.03$ in the moneyness direction and $h_2 = 0.07$ in the time to maturity direction. Single dots denote IV data obtained by inverting the BS formula separately for each observation in the sense of two-step estimators. All observations are equally weighted

4.6 Summary

In this chapter, we introduced smoothing techniques to estimate the IV smile and the IVS. We considered Nadaraya-Watson, local polynomial and least squares kernel smoothing and discussed the bandwidth choice.

In Nadaraya-Watson smoothing, one fits a local constant. This can have disadvantageous effects: due to the unequal distribution of the IV data, this may induce a bias in the wings of the smile function. In local polynomial smoothing this effect is less present. Also for larger bandwidths the bias remains small. Additionally, local polynomial smoothing allows for efficiently estimating derivatives of the regression function. This feature is ideal for estimating the LVS.

Finally, we introduced an IVS estimator based on least squares kernel smoothing. This estimator takes the option prices as input parameters, and not IV data. Thus, in computing the asymptotic confidence bands, one directly takes the nonlinear transformation of computing the IV into account.

5

Dimension-Reduced Modeling

5.1 Introduction

The IVS is a complex, high-dimensional random object. In building a model, it is thus desirable to have a low-dimensional representation of the IVS. This aim can be achieved by employing dimension reduction techniques. Generally it is found that two or three factors with appealing financial interpretations are sufficient to capture more than 90% of the IVS dynamics. This implies for instance for a scenario analysis in risk-management that only a parsimonious model needs to be implemented to study the vega-sensitivity of an option portfolio, Fengler et al. (2002b). This section will give a general overview on dimension reduction techniques in the context of IVS modeling. We will consider techniques from multivariate statistics and methods from functional data analysis. Sections 5.2 and 5.3 will provide an in-depth treatment of the CPC and the semiparametric factor model of the IVS together with an extensive empirical analysis of the German DAX index data.

In multivariate analysis, the most prominent technique for dimension reduction is principal component analysis (PCA). The idea is to seek linear combinations of the original observations, so called principal components (PCs) that inherit as much information as possible from the original data. In PCA, this means to look for standardized linear combinations with maximum variance. The approach appears to be sensible in an analysis of the IVS dynamics, since a large variance separates out systematic from idiosyncratic shocks that drive the surface. As a nice byproduct, the structure of the linear combinations reveals relationships among the variables that are not apparent in the original data. This helps understand the nature of the interdependence between different regions in the IVS.

In finance, PCA is a well-established tool in the analysis of the term structure of interest rates, see Gouriéroux et al. (1997) or Rebonato (1998) for textbook treatments: PCA is applied to a multiple time series of interest rates (or forward rates) of various maturities that is recovered from the term structure of interest rates. Typically, a small number of factors is found to represent the dynamic variations of the term structure of interest rates. The studies of Bliss (1997), Golub and Tilman (1997), Niffikeer et al. (2000), and Molgedey and Galic (2001) are examples of this kind of literature.

This approach does not immediately carry over to the analysis of IVs due to the *surface structure*. Consequently, in analogy to the interest rate case, empirical work first analyzes the term structure of IVs of ATM options, only, Zhu and Avellaneda (1997) and Fengler et al. (2002b). Alternatively, one smile at one given maturity can be analyzed within the PCA framework, Alexander (2001b). Skiadopoulos et al. (1999) group IVs into maturity buckets, average the IVs of the options, whose maturities fall into them, and apply a PCA to each bucket covariance matrix separately. A good overview of these methods can be found in Alexander (2001a).

A surface perspective on IVS dynamics is adopted in Fengler et al. (2003b) within a common principal component (CPC) framework for the IVS. The approach is motivated by two salient features that characterize the IVS dynamics: first, the instantaneous profile of the IVS is subject to changes, but most shocks tend to move it into the same direction. Second, the size of the shocks decreases with the option's maturity. This leads to high spatial correlation between contemporaneous surface values, while at the same time the 'volatility' of IV is highest for the short maturity contracts. The insight from these observations is that IVs of different maturity groups may obey a *common eigenstructure*. The CPC model exactly features this structure, since it assumes that the space spanned by the eigenvectors of the covariance matrices is identical across different groups, whereas the variances associated with the components are allowed to vary. In order to mitigate the mixing effect of IVs of different expiries, Fengler et al. (2003b) fit the daily IVS nonparametrically and investigate a number of time to maturity slices. They show that the dynamics of these slices can be generated by a small number of factors from a lower dimensional space spanned by the eigenvectors of a common transformation matrix.

Multivariate analysis is based on the idea that we observe a number of random variables on a set of objects. The interest is to study the inherent interdependence of these variables. To put the analysis of the IVS into this framework, one recovers the IVS on a grid by applying some fitting algorithm, e.g. as discussed in Sect. 4. The discrete ensemble of the observations at the grid points is treated as the set of variables. As shall be seen, this approach will yield a lot of insights into the nature of the IVS and its dynamics, and of course, it is not against the nature of multivariate analysis. However, it may be considered as being somewhat artificial, since the actual objects of interest are *functions* rather than realizations of multivariate random variables. This perspective is taken in *functional data analysis*.

In functional data analysis, we treat the observed IVS as a single entity – as a function, though discretely sampled in practice – and not as a sequence of individual observations for a choice of time to maturities and moneyness. The term 'functional' is derived from the intrinsic nature of the data, rather than from their explicit form. In treating the data in this way, the techniques of multivariate analysis can be generalized to the functional case. This leads to a functional PCA (FPCA) of the IVS as proposed by Cont and da Fonseca (2002) and Benko and Härdle (2004).

(C)PCA and also FPCA both require an estimate of the IVS. The challenge is to obtain a good fit given the degenerated string structure of the IV data. With the string structure, we recall the fact discussed already in Sect. 2.5 that in standardized markets only a very limited number of observations of the IVS exist in the time to maturity direction. Unless carefully calibrated to this structure, one may quickly obtain biased estimates. In nonparametric estimation, this will be the case when the bandwidths are chosen too big. This disadvantage is addressed in a new modeling approach by Fengler et al. (2003a). They propose a dynamic semiparametric factor (SFM) model, which approximates the IVS in a finite dimensional function space. The key feature is that this model fits in the *local* neighborhood of the design points. The approach can be considered as a combination of methods from FPCA and backfitting techniques for additive models.

In practice, functional techniques often require a discretization of the functional object that is to be estimated. Not rarely, one is back in the multivariate framework again, which additionally bears the advantages of an easy implementation and cheap computation. Also, unlike functional data analysis, the statistical properties of the techniques of multivariate analysis are usually well-known. Nevertheless, a functional approach may be considered as being more elegant. This is particularly obvious for the SFM, which delivers a biased-reduced surface estimation, dimension reduction *and* dynamic modeling in a single step.

The structure of the remaining parts of this chapter is as follows: Section 5.2 introduces the CPC models of the IVS. Since PCA is a special case of CPCA with one group only, we skip a separate presentation of PCA there. For an introduction into PCA, we refer to classical textbooks in multivariate statistics such as Mardia et al. (1992) or Härdle and Simar (2003). After a motivation of CPC models, we present their theory. Next, we derive test statistics to analyze the stability of the principal component transformation of the IVS. An empirical analysis of DAX IVs between 1995 to May 2001 follows. Section 5.3 introduces into FPCA. Section 5.4 will be devoted to the new class of SFMs. An exposition of the techniques will be given as well as an extensive empirical analysis.

128   5 Dimension-Reduced Modeling

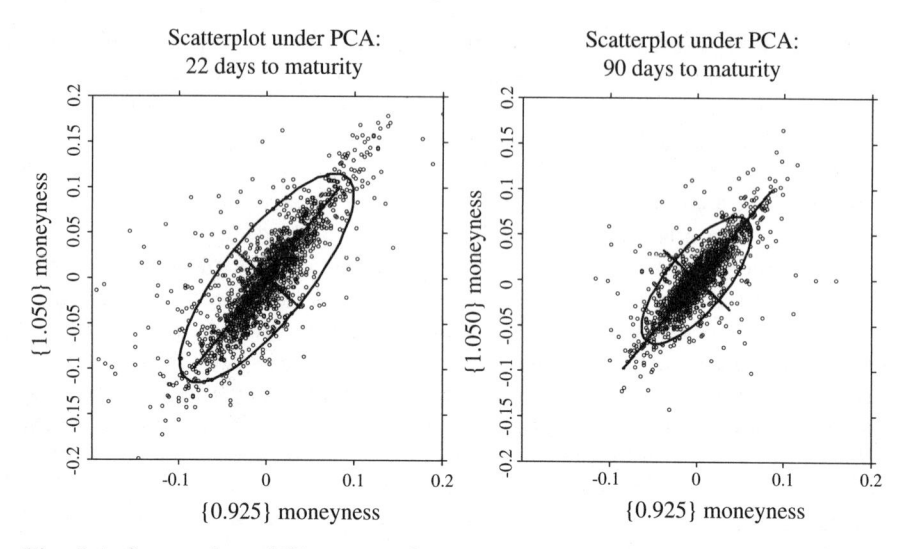

Fig. 5.1. Scatterplots of IV returns of moneyness $\kappa_f = 0.925$ against $\kappa_f = 1.050$ for the groups of 22 days and 90 days to expiry. IV returns computed as log-differences from the IVS recovered on a fixed grid. The ellipse given by the Mahalanobis distance is a 95% confidence region for a bivariate normal distribution. Principle axes of the ellipses are the eigenvectors obtained by a separate PCA for each maturity group, compare Fig. 5.2 `SCMcpcpca.xpl`

5.2 Common Principal Component Analysis

5.2.1 The Family of CPC Models

PCA, as introduced into statistics by Pearson (1901) and Hotelling (1933) is a dimension reduction technique for one group. In many applications, however, the data fall into groups in which the same variables are measured. For example, in a zoological application one measures the same characteristics across different species, Airoldi and Flury (1988), or in an economic case study one observes the same variables across different countries or markets. In an analysis of the IVS the data fall into maturity groups, as for a given observation date a limited number of maturities are traded. This is visible from the black dots in the plot of the IVS in Fig. 1.1. In these situations, it is natural to assume that the structure observed between groups is governed by one or more *common* unobservable factors. The 'degree of commonness' between the factors in each group may be of different nature. The CPC model and its related methods, which were discovered by Flury (1988), allow for a thorough analysis of the eigenstructure of the different groups.

For a graphical justification of CPC models of the IVS, observe Figs. 5.1 and 5.2: in Fig. 5.1, we present scatterplots of the IV returns of 22 days and 90 days to expiry recovered for two fixed points of (forward) moneyness. Due to the higher volatility of short term IV returns, the corresponding point cloud is

5.2 Common Principal Component Analysis

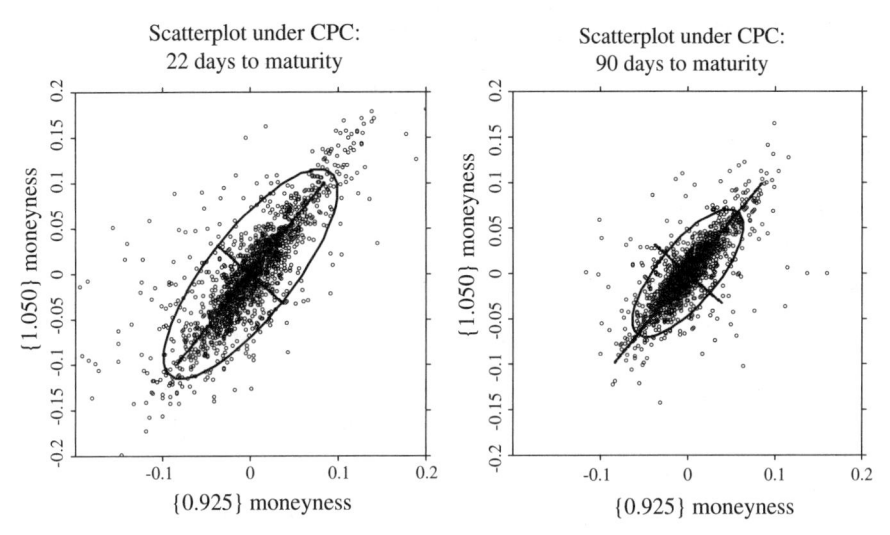

Fig. 5.2. Scatterplots of IV returns of moneyness $\kappa_f = 0.925$ against $\kappa_f = 1.050$ for the groups of 22 days and 90 days to expiry. IV returns computed as log-differences from the IVS recovered on a fixed grid. The ellipse given by the Mahalanobis distance is a 95% confidence region for a bivariate normal distribution. Principle axes of the ellipses are the eigenvectors obtained by the CPC model for *both* maturity groups, i.e. eigenvectors are estimated under the restriction to be identical, compare Fig. 5.1
SCMcpccpc.xpl

bigger compared with the one belonging to the data of the 90 days to expiry. Together with the zero mean data, we present the principal axes and the ellipse given by the Mahalanobis distance:

$$\sqrt{\mathbf{x}_i^\top \boldsymbol{\Sigma}_i^{-1} \mathbf{x}_i} = 2 \ , \ i = 1, 2 \ , \tag{5.1}$$

where $\mathbf{x}_i$ are the vectors that contain the log-differences of IVs, and $\boldsymbol{\Sigma}_i$ are the sample covariance matrices. The ellipse (5.1) is an approximate 95% confidence region for a zero mean multivariate normal distribution.

The striking observation is that the principal axes in both time to maturity groups are almost similar. It is only the *volatility* of IV that is different. A natural assumption therefore is to attribute the variability of the axes to sampling variability, and otherwise to estimate principal axes *jointly* in both groups under the constraint that they are equal: for the same data, the results are displayed in Fig. 5.2. Now, principal axes in both cases are *identical*. We shall show and test that this also holds across the short-term IVS. Consequently, via CPC methods, a significant reduction of dimension can be achieved for the IVS dynamics.

Denote by $\mathbf{X}_i \stackrel{\text{def}}{=} (\mathbf{x}_{i1}, \ldots, \mathbf{x}_{ip}) \in \mathbb{R}^p$, $i = 1, \ldots, k$ the IV returns for k maturity groups at p grid points in the IVS. The hypothesis for a CPC model is written as:

$$H_{CPC}: \quad \Psi_i = \Gamma \Lambda_i \Gamma^\top, \qquad i = 1, \ldots, k, \tag{5.2}$$

where $\Psi_1, \ldots, \Psi_k$ are positive definite $p \times p$ population covariance matrices of $\mathbf{X}_i$. Further, $\Gamma = (\gamma_1, \ldots, \gamma_p)$ denotes an orthogonal $p \times p$ matrix of eigenvectors and $\Lambda_i = \mathrm{diag}(\lambda_{i1}, \ldots, \lambda_{ip})$ is the matrix of eigenvalues. The number of parameters in the CPC model are $p(p-1)/2$ for the orthogonal matrix Γ plus kp for the eigenvalues in $\Lambda_1, \ldots, \Lambda_k$.

The PCs $\mathbf{Y}_i \stackrel{\mathrm{def}}{=} (\mathbf{y}_{i1}, \ldots, \mathbf{y}_{ip})$ are obtained by projecting $\mathbf{X}_i$ into the space spanned by its eigenvectors, i.e. by computing $\mathbf{Y}_i = \mathbf{X}_i \Gamma$. The variance of $\mathbf{Y}_i$ is

$$\mathsf{Var}(\mathbf{Y}_i) = \mathsf{Var}(\mathbf{X}_i \Gamma) = \Gamma^\top \mathsf{Var}(\mathbf{X}_i) \Gamma = \Gamma^\top \Gamma \Lambda_i \Gamma^\top \Gamma = \Lambda_i, \tag{5.3}$$

since the eigenvectors are orthogonal. This confirms that PCs are uncorrelated and that the eigenvalues correspond to their variances. The sum of the eigenvalues, i.e. $\mathrm{tr}\,\Lambda_i = \sum_j^p \lambda_{ij}$, is the total variance in the sample. If a small number of our p-variate PCs, say three of them, capture a large portion of the total variance, a considerable reduction of dimension is achieved. For we can write for each group i:

$$\mathbf{X}_i \approx \mathbf{Y}_i \widetilde{\Gamma}^\top, \tag{5.4}$$

where $\widetilde{\Gamma} = (\gamma_1, \gamma_2, \gamma_3)$. We say that the three-dimensional factor series unfolds the full set of IV returns in group i. Instead of studying a p-variate factor series, we inspect only three of them in each group. For a model in risk-management or trading, this low-dimensional series can deliver a sufficiently good description of the IVS dynamics. Furthermore, since the data in our IVS groups are very much correlated, the factor series may be regarded as *scaled versions* of each other. Thus, we can reduce our attention to study one maturity group only. In total, instead of modeling kp factor series, we end up with modeling three of them. These considerations demonstrate the usefulness of dimension reduction techniques.

A particular strength of CPC models is that they enclose a whole family of models with varying degrees of flexibility in the eigenstructure. The *proportional* model puts additional constraints on the matrix of eigenvalues Λ_i by imposing that $\lambda_{ij} = \rho_i \lambda_{1j}$, where $\rho_i > 0$ are unknown constants. This is equivalent to writing:

$$H_{prop}: \quad \Psi_i = \rho_i \Psi_1, \qquad i = 2, \ldots, k. \tag{5.5}$$

The number of parameters here are $p(p+1)/2 + (k-1)$. For the IVS this means that the variances of the common components between the groups are proportionally scaled versions of each other. In terms of modeling the IVS, this implies that one needs to resort to one maturity group only, once the scaling constants ρ_i are estimated.

In letting the eigenvalues unrestricted as in the CPC hypothesis, one can also ease the restrictions on the transformation matrix Γ: this leads to *partial* CPC models, pCPC(q), where q denotes the order of common eigenvectors in Γ. This is appropriate under the assumption that each maturity group is hit by q joint shocks, and by $(p-q)$ shocks differing among the groups. Formally, the hypothesis of the pCPC(q) model is

$$H_{pCPC} : \mathbf{\Psi}_i = \mathbf{\Gamma}^{(i)} \mathbf{\Lambda}_i \mathbf{\Gamma}^{(i)\top}, \qquad i = 1, \ldots, k, \qquad (5.6)$$

where $\mathbf{\Lambda}_i$ is as in (5.2) and $\mathbf{\Gamma}^{(i)} = \left(\mathbf{\Gamma}_c, \mathbf{\Gamma}_s^{(i)}\right)$. Here, the $p \times q$ matrix $\mathbf{\Gamma}_c$ contains the q common eigenvectors, while $\mathbf{\Gamma}_s^{(i)}$ of dimension $p \times (p-q)$ holds the $p-q$ group specific eigenvectors. The $\mathbf{\Gamma}^{(i)}$ are still orthogonal matrices. This implies that the necessary dimension to estimate a pCPC(1) model is $p = 3$. When all possible pCPC(q) are to be estimated sequentially moving from the pCPC($p-2$) down to the pCPC(1) model, it is left to the modeling approach in which order the constraints on $\boldsymbol{\gamma}_j$ are relaxed. A natural way to proceed is to allow in each step for group specific eigenvectors in the 'least important' case, where importance is measured in terms of the size of the corresponding eigenvalue. The total number of parameters amount to $p(p-1)/2 + kp + (k-1)(p-q)(p-q-1)/2$.

CPC and pCPC(q) models can be ordered in a hierarchical fashion, which allows a detailed analysis of the involved covariance matrices of different maturity groups. The highest level of similarity would be to assume equality between covariance matrices of different maturity groups $\mathbf{\Psi}_i$. In this case the number of parameters to be estimated are $p(p+1)/2$, and one may obtain the parameters by one single PCA applied to one pooled sample covariance matrix of all k groups. The models, which relax the restrictions subsequently, are the proportional model and the CPC model itself. The following levels in the hierarchy are given by the pCPC(q) models starting from $q = p - 2$ and stepping down to $q = 1$. The relations between different groups disappear subsequently, until at the last level the $\mathbf{\Psi}_i$ do not share any common eigenstructure. As all these different models are nested, one can decompose the total χ^2 statistic and test one model against a more flexible one in a *step-up procedure*. Table 5.1 displays these sequential tests. By this summation property, a test against any lower model is given by adding up the χ^2 test statistics and the degrees of freedom between the two models under comparison, Flury (1988). Additionally, we present Akaike and Schwarz information criteria for model selection.

5.2.2 Estimating Common Eigenstructures

Here, we focus on the ordinary CPC model given in (5.2) due to its practical importance and its similarity with the proportional and the pCPC models. For the theory on the other models we refer to Flury (1988).

In abuse of notation, let $\mathbf{X}_i \stackrel{\text{def}}{=} (\mathbf{x}_{i1}, \ldots, \mathbf{x}_{ip})$, $i = 1, \ldots, k$, be the $(n_i \times p)$ matrices of IV returns sampled from k underlying p-variate normal distributions $N(\boldsymbol{\mu}, \mathbf{\Psi}_i)$. As stated earlier, $\mathbf{\Psi}_i$ denotes the population covariance matrix. In our view the sample is recovered from a grid of size $(k \times p)$ obtained by smoothing the IVS as discussed in the previous chapter. Let $\mathbf{\Sigma}_i$ be

5 Dimension-Reduced Modeling

Table 5.1. The table presents the hierarchy of nested CPC models. From top to bottom restrictions on the estimated population covariance matrices are eased. Sequentially, starting from top, each model is tested against the next lower one in the hierarchy. The degrees of freedom of the corresponding χ^2 test as given in column (3) are obtained by subtracting the number of parameters to be estimated in each model, compare Flury (1988), p. 151, and Fengler et al. (2003b). After arriving at the CPC hypothesis, one tests the CPC against the pCPC$(p-2)$ model. Next, the pCPC$(p-2)$ model is tested against the pCPC$(p-3)$ model, and so on, down to the pCPC(1) model which is finally tested against the hypothesis of arbitrary covariance matrices

Higher Model	Lower Model	Degrees of Freedom
equality	proportionality	$k-1$
proportionality	CPC	$(p-1)(k-1)$
CPC	pCPC(q) $(1 \leq q \leq p-2)$	$\frac{1}{2}(k-1)(p-q)(p-q-1)$
pCPC(1)	arbitrary covariance matrices	$(p-1)(k-1)$

the (unbiased) sample covariance matrix of the returns of IV. In our applications, we derive returns as first order log-differences of IVs. The sample size is $n_i > p$ for $i = 1, \ldots, k$.

Applying to general results from multivariate analysis, Härdle and Simar (2003), under the assumption of normality, the distribution of Σ_i is a generalization of the chi-squared variate, the *Wishart distribution* with scale matrix Ψ_i and $(n_i - 1)$ degrees of freedom. It is denoted by:

$$n_i \Sigma_i \sim W_p(\Psi_i, n_i - 1) \ .$$

The pdf of the Wishart distribution is:

$$f(\Sigma) = \frac{1}{\Gamma_p(\frac{n-1}{2})|\Psi|^{(n-1)/2}} \left(\frac{n-1}{2}\right)^{p(n-1)/2}$$
$$\times \ \exp \operatorname{tr}\left(-\frac{n-1}{2}\Psi^{-1}\Sigma\right)|\Sigma|^{(n-p-2)/2} \quad (5.7)$$

for Σ positive definite, and zero otherwise, Evans et al. (2000).

$$\Gamma_p(u) \stackrel{\text{def}}{=} \pi^{p(p-1)/4} \prod_{j=1}^{p} \Gamma\left\{u - \frac{1}{2}(j-1)\right\} \quad (5.8)$$

denotes the multivariate Gamma function, where $\pi = 3.141...$, and $\Gamma(t) \stackrel{\text{def}}{=} \int_0^\infty e^{-s} s^{t-1} \, ds$ is the univariate Gamma function.

For the k Wishart matrices Σ_i the likelihood function is given by

$$L(\Psi_1, \ldots, \Psi_k) = c \prod_{i=1}^{k} \exp\left[\operatorname{tr}\left\{-\frac{1}{2}(n_i - 1)\Psi_i^{-1}\Sigma_i\right\}\right] |\Psi_i|^{-\frac{1}{2}(n_i - 1)} , \quad (5.9)$$

5.2 Common Principal Component Analysis

where c is a constant not depending on the parameters. Maximizing the likelihood is equivalent to minimizing the function

$$g(\mathbf{\Gamma}, \mathbf{\Psi}_1, \ldots, \mathbf{\Psi}_k) = \sum_{i=1}^{k}(n_i - 1)\left\{\ln |\mathbf{\Psi}_i| + \text{tr}(\mathbf{\Psi}_i^{-1}\mathbf{\Sigma}_i)\right\}. \tag{5.10}$$

Assuming that H_{CPC} in (5.2) holds, yields

$$g(\mathbf{\Gamma}, \mathbf{\Lambda}_1, \ldots, \mathbf{\Lambda}_k) = \sum_{i=1}^{k}(n_i - 1)\sum_{j=1}^{p}\left(\ln \lambda_{ij} + \frac{\boldsymbol{\gamma}_j^\top \mathbf{\Sigma}_i \boldsymbol{\gamma}_j}{\lambda_{ij}}\right). \tag{5.11}$$

We impose the orthogonality constraints of $\mathbf{\Gamma}$ by introducing the Lagrange multiplyers μ_j for the p constraints $\boldsymbol{\gamma}_j^\top \boldsymbol{\gamma}_j = 1$, and the Lagrange multiplyers μ_{hj} for the $p(p-1)/2$ constraints $\boldsymbol{\gamma}_h^\top \boldsymbol{\gamma}_j = 0$ $(h \neq j)$. Hence the Lagrange function to be minimized is given by

$$g^*(\mathbf{\Gamma}, \mathbf{\Lambda}_1, \ldots, \mathbf{\Lambda}_k) = g(\cdot) - \sum_{j=1}^{p}\mu_j(\boldsymbol{\gamma}_j^\top \boldsymbol{\gamma}_j - 1) - 2\sum_{h<j}^{p}\mu_{hj}\boldsymbol{\gamma}_h^\top \boldsymbol{\gamma}_j. \tag{5.12}$$

Taking partial derivatives with respect to all λ_{ij} and $\boldsymbol{\gamma}_j$, it can be shown that the solution of the CPC model can be written as the generalized system of characteristic equations:

$$\boldsymbol{\gamma}_m^\top \left\{\sum_{i=1}^{k}(n_i - 1)\frac{\lambda_{im} - \lambda_{ij}}{\lambda_{im}\lambda_{ij}}\mathbf{\Sigma}_i\right\}\boldsymbol{\gamma}_j = 0, \quad m, j = 1, \ldots, p, \quad m \neq j. \tag{5.13}$$

This is solved observing

$$\lambda_{im} = \boldsymbol{\gamma}_m^\top \mathbf{\Sigma}_i \boldsymbol{\gamma}_m, \quad i = 1, \ldots, k, \quad m = 1, \ldots, p, \tag{5.14}$$

and the constraints:

$$\boldsymbol{\gamma}_m^\top \boldsymbol{\gamma}_j = \begin{cases} 0 & m \neq j \\ 1 & m = j \end{cases}. \tag{5.15}$$

If $k = 1$, the one-group case, it is quickly seen that (5.13) to (5.15) collapse to the usual system of equations for an eigenvalue problem of $\mathbf{\Sigma}$; this leads to an ordinary PCA.

Flury (1988) proves existence and uniqueness of the maximum of the likelihood function, and Flury and Gautschi (1986) provide a numerical algorithm solving (5.13) to (5.15) that is implemented in XploRe, Härdle et al. (2000b). The maximum likelihood estimates of $\mathbf{\Psi}_i$ are denoted by $\hat{\mathbf{\Psi}}_i = \hat{\mathbf{\Gamma}}\hat{\mathbf{\Lambda}}_i\hat{\mathbf{\Gamma}}^\top$, $i = 1, \ldots, k$. Sample common PCs of the maturity groups are given by $\mathbf{Y}_i = \mathbf{X}_i\hat{\mathbf{\Gamma}}$.

Furthermore the following results are due to Flury (1988):

The estimated eigenvalues $\hat{\lambda}_{ij}$, $i = 1, \ldots, k$, $j = 1, \ldots, p$ are asymptotically distributed as

$$\sqrt{n_i - 1}(\hat{\lambda}_{ij} - \lambda_{ij}) \xrightarrow{\mathcal{L}} N(0, 2\lambda_{ij}^2) \qquad (5.16)$$

as $\min n_i \uparrow \infty$, and are asymptotically independent of each other and independent of $\hat{\Gamma}$.

Denote by $N = \sum_{i=1}^{k} n_i$ the overall number of observations. The asymptotic distribution of the p eigenvectors is given by:

$$\sqrt{N - k}\left(\hat{\Gamma} - \Gamma\right) \xrightarrow{\mathcal{L}} N\left(0, \text{Var}(\Gamma)\right), \qquad (5.17)$$

where $\text{Var}(\Gamma)$ is the $p^2 \times p^2$ matrix

$$\text{Var}(\Gamma) = \begin{pmatrix} \sum_{\substack{j=1 \\ j \neq 1}}^{p} \theta_{1j}\gamma_j\gamma_j^\top & -\theta_{12}\gamma_2\gamma_1^\top & \cdots & -\theta_{1p}\gamma_p\gamma_1^\top \\ -\theta_{21}\gamma_1\gamma_2^\top & \sum_{\substack{j=1 \\ j \neq 2}}^{p} \theta_{1j}\gamma_j\gamma_j^\top & \cdots & -\theta_{2p}\gamma_p\gamma_1^\top \\ \vdots & \vdots & \ddots & \vdots \\ -\theta_{p1}\gamma_1\gamma_p^\top & -\theta_{p2}\gamma_2\gamma_p^\top & \cdots & \sum_{\substack{j=1 \\ j \neq p}}^{p} \theta_{pj}\gamma_j\gamma_j^\top \end{pmatrix} \qquad (5.18)$$

and $\theta_{jm} \stackrel{\text{def}}{=} \left\{\sum_{i=1}^{k}\left(\frac{N-k}{n_i-1}\frac{\lambda_{ij}\lambda_{im}}{(\lambda_{ij}-\lambda_{im})^2}\right)^{-1}\right\}^{-1}$ with $m \neq j$. We point out that the variance matrix, as usual in PCA, does *not* have full rank. Instead, it has rank $p(p-1)/2$.

5.2.3 Stability Tests for Eigenvalues and Eigenvectors

With the preceding results, we are in the shape of conducting hypothesis tests about the eigenvalues λ_{ij} and the eigenvectors γ_j in the multisample framework. The tests are formulated in this section.

Eigenvalues

Suppose we estimate a CPC model in R subsamples and wish to test the hypothesis of equality of the jth eigenvalue $\lambda_{ij}^{(r)}$ in the ith group across R subsamples:

$$H_0: \quad \lambda_{ij}^{(1)} = \cdots = \lambda_{ij}^{(r)} = \cdots = \lambda_{ij}^{(R)}$$

against the alternative $H_1: \exists\, \lambda_{ij}^{(r_1)}, \lambda_{ij}^{(r_2)}$ such that $\lambda_{ij}^{(r_1)} \neq \lambda_{ij}^{(r_2)}$ for some r_1, r_2. H_0 can be written as

5.2 Common Principal Component Analysis

$$H_0: \begin{matrix} \lambda_{ij}^{(1)} - \lambda_{ij}^{(2)} = 0 \\ \vdots \\ \lambda_{ij}^{(1)} - \lambda_{ij}^{(r)} = 0 \\ \vdots \\ \lambda_{ij}^{(1)} - \lambda_{ij}^{(R)} = 0 \end{matrix} \quad . \tag{5.19}$$

To formulate the test statistic, it is useful to define a contrast matrix: $\mathbf{C} = (\mathbf{c}_1, \ldots, \mathbf{c}_L)$ is called a contrast matrix, if $\sum_{l=1}^{L} \mathbf{c}_l = \mathbf{0}$, and if its rows are linearly independent, Johnson and Wichern (1998).

Especially, define by $\mathbf{C}_1$ the $(R-1) \times R$ contrast matrix

$$\mathbf{C}_1 = \begin{pmatrix} 1 & -1 & 0 & \cdots & 0 \\ 1 & 0 & -1 & \cdots & 0 \\ \vdots & \vdots & \vdots & \ddots & \vdots \\ 1 & 0 & 0 & \cdots & -1 \end{pmatrix} . \tag{5.20}$$

Equality of the jth eigenvalue in the ith group in R subsamples.
Denote by $\widetilde{\boldsymbol{\lambda}}_{ij}$ the $R \times 1$ stacked vector of $\hat{\lambda}_{ij}^{(r)}$, in $r = 1, \ldots, R$ subsamples and its asymptotic variance by the $R \times R$ matrix

$$\mathsf{Var}(\widetilde{\boldsymbol{\lambda}}) = 2 \; \mathsf{diag} \left(\frac{\lambda_{ij}^{(1)\,2}}{n_{i1} - 1}, \ldots, \frac{\lambda_{ij}^{(r)\,2}}{n_{ir} - 1}, \ldots, \frac{\lambda_{ij}^{(R)\,2}}{n_{iR} - 1} \right),$$

where n_{ir} is the sample size of group i and subsample r. A test for (5.19) can be based on:

$$T_{\text{equ}} = (\mathbf{C}_1 \widetilde{\boldsymbol{\lambda}}_{ij})^\top \left\{ \mathbf{C}_1 \mathsf{Var}(\widetilde{\boldsymbol{\lambda}}) \mathbf{C}_1^\top \right\}^{-1} \mathbf{C}_1 \widetilde{\boldsymbol{\lambda}}_{ij} . \tag{5.21}$$

Since the λ_{ij} are asymptotically normal and independent by virtue of (5.16), $\mathbf{z} \stackrel{\text{def}}{=} \left\{ \mathbf{C}_1 \mathsf{Var}(\widetilde{\boldsymbol{\lambda}}) \mathbf{C}_1^\top \right\}^{-1/2} \mathbf{C}_1 \widetilde{\boldsymbol{\lambda}}_{ij}$ is asymptotically $N(\mathbf{0}_{R-1}, \mathbf{I}_{R-1})$ under H_0. Thus $T_{\text{equ}} = \mathbf{z}^\top \mathbf{z}$ is asymptotically χ^2 distributed with $(R-1)$ degrees of freedom. In practice all unknowns are to be replaced by consistent estimates, which does not alter the asymptotic distribution of (5.21).

In fact, there are a lot of ways of formulating the aforementioned hypothesis by a different choice of the contrast matrix. However, the test statistic does not depend on this particular choice. For example, an equivalent formulation of the hypothesis using the contrast matrix

$$\mathbf{C}_2 = \begin{pmatrix} -1 & 1 & 0 & \cdots & 0 & 0 \\ 0 & -1 & 1 & \cdots & 0 & 0 \\ \vdots & \vdots & \vdots & \ddots & \vdots & \vdots \\ 0 & 0 & 0 & \cdots & -1 & 1 \end{pmatrix}$$

would be

$$H_0: \begin{aligned} \lambda_{ij}^{(2)} - \lambda_{ij}^{(1)} &= 0 \\ \lambda_{ij}^{(3)} - \lambda_{ij}^{(2)} &= 0 \\ &\vdots \\ \lambda_{ij}^{(R)} - \lambda_{ij}^{(R-1)} &= 0 \end{aligned}.$$

The equivalence of the tests is due to the fact that any pair of contrast matrices is related by a nonsingular matrix $\mathbf{A}$ such that $\mathbf{C}_1 = \mathbf{A}\mathbf{C}_2$. Inserting $\mathbf{A}\mathbf{C}_2$ into T yields:

$$\begin{aligned} T &= (\mathbf{C}_1\tilde{\boldsymbol{\lambda}})^\top \{\mathbf{C}_1 \mathsf{Var}(\tilde{\boldsymbol{\lambda}})\mathbf{C}_1^\top\}^{-1} \mathbf{C}_1\tilde{\boldsymbol{\lambda}} \\ &= (\mathbf{A}\mathbf{C}_2\tilde{\boldsymbol{\lambda}})^\top \{\mathbf{A}\mathbf{C}_2 \mathsf{Var}(\tilde{\boldsymbol{\lambda}})\mathbf{C}_2^\top \mathbf{A}^\top\}^{-1} \mathbf{A}\mathbf{C}_2\tilde{\boldsymbol{\lambda}} \\ &= (\mathbf{C}_2\tilde{\boldsymbol{\lambda}})^\top \{\mathbf{C}_2 \mathsf{Var}(\tilde{\boldsymbol{\lambda}})\mathbf{C}_2^\top\}^{-1} \mathbf{C}_2\tilde{\boldsymbol{\lambda}} \ , \end{aligned}$$

which is the same as before.

Eigenvectors

In testing for the eigenvectors one faces the difficulty that the covariance matrix $\mathsf{Var}(\boldsymbol{\Gamma})$ given in (5.18) is singular. The problem was first solved for testing one single eigenvector by Anderson (1963) and generalized for the case of several eigenvectors by Flury (1988). We adapt their strategies here for our stability tests.

We will be interested in testing for stability of a single eigenvector across different samples. Without loss of generality we focus on the first eigenvector. Thus, the test will be based on the upper $p \times p$ matrix $\sum_{\substack{j=1 \\ j \neq 1}}^{p} \theta_{1j} \boldsymbol{\gamma}_j \boldsymbol{\gamma}_j^\top$ in (5.18), only. Tests for equality of $q > 1$ eigenvectors would need to employ the $qp \times qp$ upper submatrix of (5.18).

In analogy to (5.19) write

$$H_0: \boldsymbol{\gamma}_1^{(1)} = \cdots = \boldsymbol{\gamma}_1^{(r)} = \cdots = \boldsymbol{\gamma}_1^{(R)}$$

against the alternative $H_1: \exists\ \boldsymbol{\gamma}_1^{(r_1)}, \boldsymbol{\gamma}_1^{(r_2)}$ such that $\boldsymbol{\gamma}_1^{(r_1)} \neq \boldsymbol{\gamma}_1^{(r_2)}$ for some r_1, r_2.

Again we rewrite H_0 as

$$H_0: \begin{aligned} \boldsymbol{\gamma}_1^{(1)} - \boldsymbol{\gamma}_1^{(2)} &= \mathbf{0}_p \\ \boldsymbol{\gamma}_1^{(1)} - \boldsymbol{\gamma}_1^{(3)} &= \mathbf{0}_p \\ &\vdots \\ \boldsymbol{\gamma}_1^{(1)} - \boldsymbol{\gamma}_1^{(r)} &= \mathbf{0}_p \\ &\vdots \\ \boldsymbol{\gamma}_1^{(1)} - \boldsymbol{\gamma}_1^{(R)} &= \mathbf{0}_p \end{aligned}, \quad (5.22)$$

and use the $p(R-1) \times pR$ contrast matrix:

$$\mathbf{C}_3 = \begin{pmatrix} \mathbf{I} & -\mathbf{I} & \mathbf{0} & \cdots & \mathbf{0} \\ \mathbf{I} & \mathbf{0} & -\mathbf{I} & \cdots & \mathbf{0} \\ \vdots & \vdots & \vdots & \ddots & \vdots \\ \mathbf{I} & \mathbf{0} & \mathbf{0} & \cdots & -\mathbf{I} \end{pmatrix}, \quad (5.23)$$

where $\mathbf{I} \stackrel{\text{def}}{=} \mathbf{I}_p$ and $\mathbf{0} \stackrel{\text{def}}{=} \mathbf{0}_{p \times p}$, here.

Test of equality of the first eigenvector in R subsamples. Denote by $\widetilde{\boldsymbol{\gamma}}_1$ the $pR \times 1$ vector of stacked eigenvectors $\boldsymbol{\gamma}_1^{(r)}$. Suppose that the R subsamples are independently drawn, and define $\mathsf{Var}(\widetilde{\boldsymbol{\gamma}})$ as the $pR \times pR$ block-diagonal matrix

$$\mathsf{Var}(\widetilde{\boldsymbol{\gamma}}) = \begin{pmatrix} \sum_{\substack{j=1 \\ j \neq 1}}^{p} \theta_{1j} \boldsymbol{\gamma}_j^{(1)} \boldsymbol{\gamma}_j^{(1)\top} & \cdots & 0 \\ \vdots & \ddots & \vdots \\ 0 & \cdots & \sum_{\substack{j=1 \\ j \neq 1}}^{p} \theta_{1j} \boldsymbol{\gamma}_j^{(R)} \boldsymbol{\gamma}_j^{(R)\top} \end{pmatrix}. \quad (5.24)$$

The α-level test of equality

$$H_0 : \mathbf{C}_3 \widetilde{\boldsymbol{\gamma}}_1 = \mathbf{0}_{R(p-1)}$$

for the R first eigenvectors against $H_1 : \mathbf{C}_3 \widetilde{\boldsymbol{\gamma}}_1 \neq 0$ is given by:

$$T_{\text{equ}} = (\mathbf{C}_3 \widetilde{\boldsymbol{\gamma}}_1)^\top \{\mathbf{C}_3 \mathsf{Var}(\widetilde{\boldsymbol{\gamma}}) \mathbf{C}_3^\top\}^{-1} \mathbf{C}_3 \widetilde{\boldsymbol{\gamma}}_1. \quad (5.25)$$

Since $\sum_{\substack{j=1 \\ j \neq 1}}^{p} \theta_{1j} \boldsymbol{\gamma}_j \boldsymbol{\gamma}_j^\top$ has rank $p-1$, the $pR \times pR$ matrix $\mathsf{Var}(\widetilde{\boldsymbol{\gamma}})$ has rank $R(p-1)$. Thus $\mathbf{C}_3 \mathsf{Var}(\widetilde{\boldsymbol{\gamma}}) \mathbf{C}_3^\top$ has full rank only if $R(p-1) \geq (R-1)p$, or, equivalently, when $p \geq R$. In this case, since the $\boldsymbol{\gamma}_j$ are asymptotically normal by (5.17), $\mathbf{z} \stackrel{\text{def}}{=} \{\mathbf{C}_3 \mathsf{Var}(\widetilde{\boldsymbol{\gamma}}) \mathbf{C}_3^\top\}^{-1/2} \mathbf{C}_3 \widetilde{\boldsymbol{\gamma}}_1$ is asymptotically $N(\mathbf{0}_{p(R-1)}, \mathbf{I}_{p(R-1)})$ under H_0. Thus $T_{\text{equ}} = \mathbf{z}^\top \mathbf{z}$ is asymptotically χ^2 distributed with $p(R-1)$ degrees of freedom.

If $p < R$, the test is computed with the generalized inverse of $\mathbf{C}_3 \mathsf{Var}(\widetilde{\boldsymbol{\gamma}}) \mathbf{C}_3^\top$. Then, by Theorem 1 in Khatri (1980) on quadratic forms of (singular) normal variables, $T_{\text{equ}} = \mathbf{z}^\top \mathbf{z}$ is asymptotically χ^2 distributed with $(p-1)R$ degrees of freedom.

Thus, H_0 is rejected, if

$$T_{\text{equ}} > \chi^2\left(1 - \alpha; \min\{(p-1)R, p(R-1)\}\right), \quad (5.26)$$

where $\chi^2(1-\alpha; \nu)$, is the $(1-\alpha)$-quantile of the chi-squared distribution with ν degrees of freedom.

5.2.4 CPC Model Selection

There are several strategies of model selection. Given our maximum likelihood framework, on the one hand, one may construct likelihood ratio tests and test each model separately against the unrestricted model. The log-likelihood ratio statistic for testing the H_{CPC} against the unrestricted model (unrelatedness between covariance matrices) is given by:

$$T = -2\ln\frac{L(\hat{\boldsymbol{\Psi}}_1,\ldots,\hat{\boldsymbol{\Psi}}_k)}{L(\boldsymbol{S}_1,\ldots,\boldsymbol{S}_k)} = \sum_{i=1}^{k}(n_i-1)\ln\frac{|\hat{\boldsymbol{\Psi}}_i|}{|\boldsymbol{S}_i|}, \quad (5.27)$$

where $L(\boldsymbol{S}_1,\ldots,\boldsymbol{S}_k)$ denotes the unrestricted maximum of the log-likelihood function. The number of parameters estimated in the CPC model are $p(p-1)/2$ for the orthogonal matrix $\boldsymbol{\Gamma}$ plus kp for the eigenvalues $\boldsymbol{\Lambda}_i$, and the number of parameters in the unrelated case are given by $kp(p-1)/2 + kp$. Hence the test is asymptotically chi-squared with $(k-1)p(p-1)/2$ degrees of freedom as $\min n_i \uparrow \infty$, Rao (1973).

On the other hand, it was said that the CPC models are nested, since each model implies all the models which are lower in the hierarchy. For instance, the proportional model necessarily implies the CPC model, or a pCPC(3) model implies the pCPC(2) model. From this feature, one can decompose the total chi-squared statistic, i.e. the test of equality against inequality, into partial chi-squared statistics in the following way, Flury (1988):

$$T_{total} = T(\text{inequality of proportionality constants} \mid \text{proportionality})$$
$$+ T(\text{deviation from proportionality} \mid \text{CPC})$$
$$+ T(\text{nonequality of last } p-q \text{ components} \mid \text{pCPC}(q))$$
$$+ T(\text{nonequality of the first } q \text{ components}).$$

This decomposition of the log-likelihood function and testing along these lines, i.e. the more restrictive model against the less restrictive model, is called *step-up procedure*. The degrees of freedom of these sequential tests have already been presented in Table 5.1.

Alternative model selection approaches are the AIC and SIC criteria, Akaike (1973) and Schwarz (1978), see also our discussions on bandwidth choice in Sect. 4.4 and 5.4.3. The AIC is defined by:

$$\Xi_{AIC} \stackrel{\text{def}}{=} -2 \times (\text{maximum of log-likelihood})$$
$$+ 2 \times (\text{number of parameters estimated}).$$

Following Flury (1988), we use a modified AIC. Assume we have U hierarchically ordered models to compare, with $a_1 < \ldots < a_u < \ldots < a_U$ parameters in model u. Then define the modified AIC as:

$$\Xi_{AIC}(u) \stackrel{\text{def}}{=} -2(L_u - L_U) + 2(a_u - a_1), \quad (5.28)$$

where L_u is the maximum of the log-likelihood function of model u. Selecting the model with the lowest ordinary AIC is equivalent to selecting the model with the lowest modified $\Xi_{AIC}(u)$. Observe that $\Xi_{AIC}(U) = 2(a_U - a_1)$ and $\Xi_{AIC}(1) = -2(L_1 - L_U)$.

The SIC, which aims at finite dimensional models, is defined as

$$\Xi_{SIC} \stackrel{\text{def}}{=} -2 \times (\text{maximum of log-likelihood}) \\ + (\text{number of parameters}) \times \ln(\text{number of observations}).$$

As in Fengler et al. (2003b), we modify this criterion to:

$$\Xi_{SIC}(u) \stackrel{\text{def}}{=} -2(L_u - L_U) + (a_u - a_1)\ln(N), \tag{5.29}$$

where $N = \sum_{i=1}^{k} n_i$ denotes the overall number of observations across the k groups. The model with the lowest SIC is the best fitting one.

5.2.5 Empirical Results

For the empirical CPC analysis, we estimate the IVS for the 1995 to 2001 data from daily samples by means of a local polynomial estimator, Sect. 4.3. The data set is described in the appendix. The moneyness grid is $\kappa_f \in \{0.925, 0.950, 0.975, 1.000, 1.025, 1.050\}$ and the maturity grid is $\tau \in \{0.0625, 0.1250, 0.1875, 0.2500\}$ years, which corresponds to 22, 45, 68, and 90 days to expiry. As kernel function we choose the product of univariate quartic kernels. In the bandwidth selection, we proceed as discussed in Sect. 4.4.2. Since our estimation grid only covers the short maturity data, there is no particular need for a localization of bandwidths. For robustness, IVs with maturity of less than 10 days are excluded from the estimation.

First, we will estimate the family of CPC models in the entire sample period. This will be followed by a stability analysis of eigenvalues and eigenvectors across the different samples.

The Entire Sample Period

The results of our model selection procedures are displayed in Table 5.2. According to the sequential chi-squared tests the model to be preferred is a pCPC(1) model, since this test is the first that cannot be rejected against the next more flexible model. Also when testing directly against the unrelated model, which is done by adding up the test statistics and the corresponding degrees of freedom between the model of interest and the unrelated model in Table 5.2, it is the pCPC(1) model which is not rejected. AIC and SIC both recommend the pCPC(1). Note also that according to the SIC all CPC(q) models with $q \leq 3$ are superior to the unrelated model. For the remaining CPC models, the SIC is slightly higher than for the unrelated case, whereas for the proportional and the equality models the information criteria increase

5 Dimension-Reduced Modeling

Table 5.2. Step-up & model building approach of CPC models

Model Higher	Lower	Chi. Sqr.	df.	p-val.	AIC	SIC
Equality	Proportionality	1174.9	3	0.00	3529.3	3529.3
Proportionality	CPC	1488.8	15	0.00	2360.3	2407.0
CPC	pCPC(4)	122.6	3	0.00	901.5	1181.2
pCPC(4)	pCPC(3)	210.6	6	0.00	784.9	1111.2
pCPC(3)	pCPC(2)	115.5	9	0.00	586.2	1005.8
pCPC(2)	pCPC(1)	398.9	12	0.00	488.7	1048.2
pCPC(1)	Unrelated	17.6	15	0.28	113.7	859.7
Unrelated					126.0	1105.1

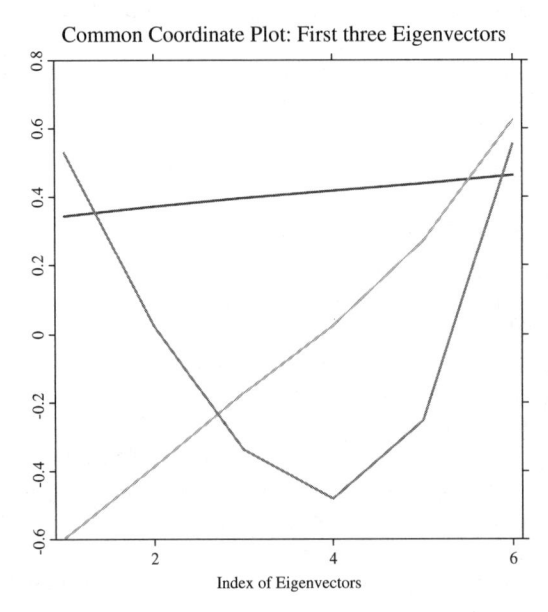

Fig. 5.3. CPC model for the entire sample period 19950101–20010531. First eigenvector horizontal line, second eigenvector diagonal line, third eigenvector U-shaped line. Compare with Fengler et al. (2003b)

tremendously. As shall be seen in the following, for an approximation up to 88%, one component will be sufficient, while the second and third only add only 6% and 3% of explained variance. Thus, we believe – also for computational and practical simplicity – that a CPC model can be chosen as a valid description of the IVS dynamics.

The estimation results of the eigenvectors for the entire sample period exhibit the same stylized facts as documented in Fengler et al. (2003b) for the year 1999 for daily settlement prices. In Fig. 5.3, we display the results for the first three eigenvectors. Table 5.3 reports the estimation results of the entire

5.2 Common Principal Component Analysis

Table 5.3. In the top position the eigenvectors $\hat{\boldsymbol{\Gamma}} = (\hat{\boldsymbol{\gamma}}_1, \ldots, \hat{\boldsymbol{\gamma}}_6)$. From top to bottom, the numbers denote the moneyness grid $\kappa_f \in \{0.925, 0.950, 0.975, 1.000, 1.025, 1.050\}$. The eigenvalues below $\hat{\lambda}_{ij} \times 10^3$ are ordered from top to bottom with increasing maturity $\tau_i \in \{0.0625, 0.1250, 0.1875, 0.2500\}$, standard errors in parenthesis; sample period 19950101 to 20010531

mon. index	$\hat{\gamma}_1$	$\hat{\gamma}_2$	$\hat{\gamma}_3$	$\hat{\gamma}_4$	$\hat{\gamma}_5$	$\hat{\gamma}_6$
1	0.344	−0.598	0.530	0.472	−0.129	0.055
	(0.0021)	(0.0095)	(0.0113)	(0.0070)	(0.0085)	(0.0037)
2	0.373	−0.385	0.022	−0.614	0.502	−0.288
	(0.0014)	(0.0044)	(0.0096)	(0.0086)	(0.0105)	(0.0068)
3	0.397	−0.173	−0.339	−0.326	−0.457	0.618
	(0.0010)	(0.0065)	(0.0055)	(0.0090)	(0.0090)	(0.0056)
4	0.419	0.024	−0.482	0.250	−0.337	−0.644
	(0.0011)	(0.0085)	(0.0038)	(0.0083)	(0.0088)	(0.0043)
5	0.440	0.270	−0.252	0.432	0.610	0.334
	(0.0012)	(0.0057)	(0.0074)	(0.0105)	(0.0081)	(0.0074)
6	0.463	0.625	0.554	−0.213	−0.191	−0.073
	(0.0022)	(0.0095)	(0.0108)	(0.0076)	(0.0052)	(0.0032)
mat. group	$\hat{\lambda}_{i1}$	$\hat{\lambda}_{i2}$	$\hat{\lambda}_{i3}$	$\hat{\lambda}_{i4}$	$\hat{\lambda}_{i5}$	$\hat{\lambda}_{i6}$
1	16.39	0.90	0.55	0.11	0.04	0.01
	(0.578)	(0.032)	(0.019)	(0.004)	(0.001)	(0.0003)
2	10.14	0.41	0.16	0.07	0.03	0.01
	(0.357)	(0.014)	(0.006)	(0.002)	(0.001)	(0.0004)
3	7.20	0.33	0.14	0.07	0.04	0.02
	(0.254)	(0.012)	(0.005)	(0.002)	(0.001)	(0.001)
4	6.01	0.40	0.23	0.09	0.06	0.02
	(0.211)	(0.014)	(0.008)	(0.003)	(0.002)	(0.001)

matrix of eigenvectors. The numbers given in parenthesis are the asymptotic standard errors.

The factor loadings of the first eigenvector, the blue line in Fig. 5.3, are of the same sign throughout (eigenvectors are unique up to sign), and give approximately the same weight to each volatility shock across the smile. We hence interpret this factor as a *common shift* factor. In Fig. 5.4, we present the projection of the longest IV maturity group (three months maturity) using the first eigenvector. The upper panel shows the PC, the lower the integrated process. The shift interpretation of the first component is also visible from the general structure of this process: in comparison with Fig. 2.13, it is seen that it exhibits almost the same patterns as the IV process itself.

In PCA, one typically employs the following measure to gauge the fraction of variance, which is captured by the j'th factor:

$$\frac{\hat{\lambda}_{ij'}}{\sum_{j=1}^{p}\hat{\lambda}_{ij}}, \qquad (5.30)$$

142   5 Dimension-Reduced Modeling

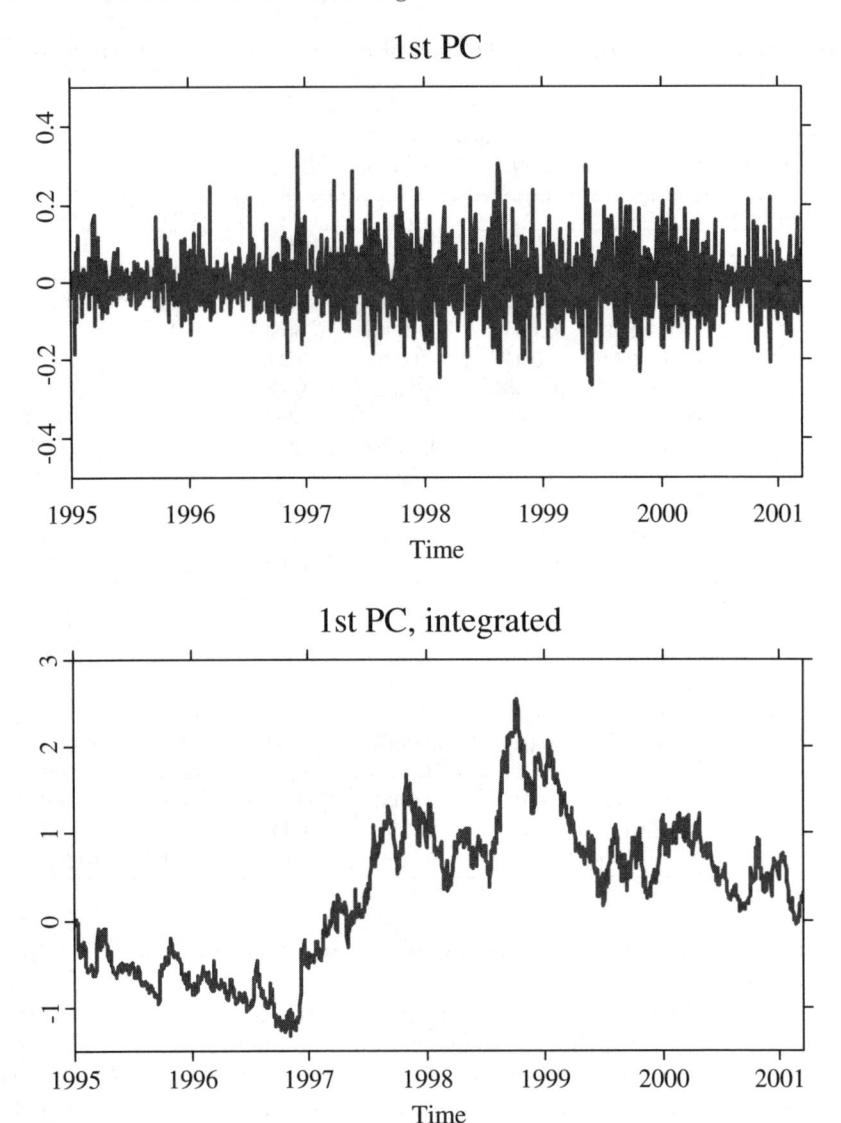

1st PC

1st PC, integrated

Fig. 5.4. Projection of the longest maturity group (90 days to expiry) using the first eigenvector. The *upper panel* shows returns, the *lower panel* the integrated series

for $i = 1, \ldots, k$ and $j = 1, \ldots, p$. This is reasonable, since PCs are uncorrelated by construction and each eigenvalue is the variance of the corresponding PC. For the first PC, this amounts to 88% in the longest maturity group.

The second eigenvector, the green line, switches its sign at ATM, and gives opposite weights to the shocks in the wings of the IVS. Thus, we interpret the second type of shocks as *common slope* shocks. Figure 5.5 displays this

2nd PC

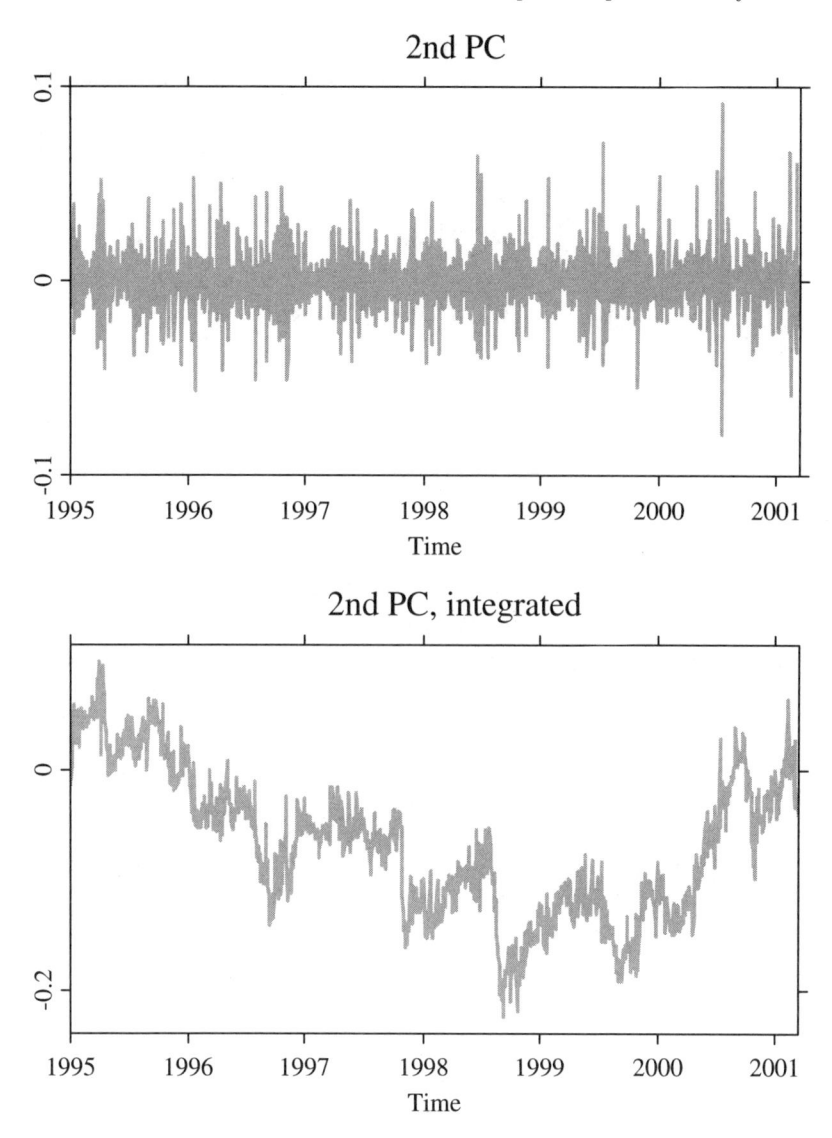

2nd PC, integrated

Fig. 5.5. Projection of the longest maturity group (90 days to expiry) using the second eigenvector. The *upper panel* shows returns, the *lower panel* the integrated series

component. The integrated second PC has a stable downward trend, which appears to revert around 1999. The third eigenvector can be interpreted as a *common twist* factor. This factor hits the curvature of the surface, since the sign of the eigenvector switches within the near-the-money region. Again the projection and the integrated process are shown in Fig. 5.6. These components account for only 6% and 3% of the variance. Similar results have been obtained

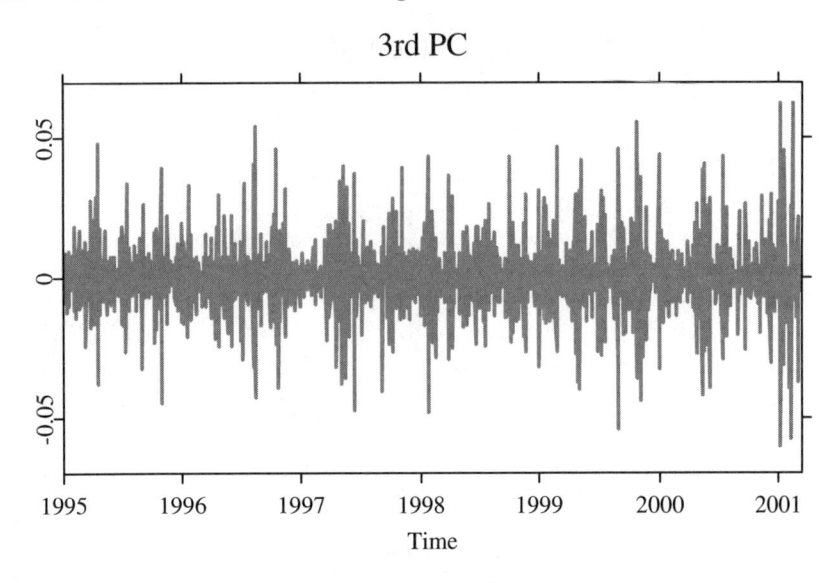

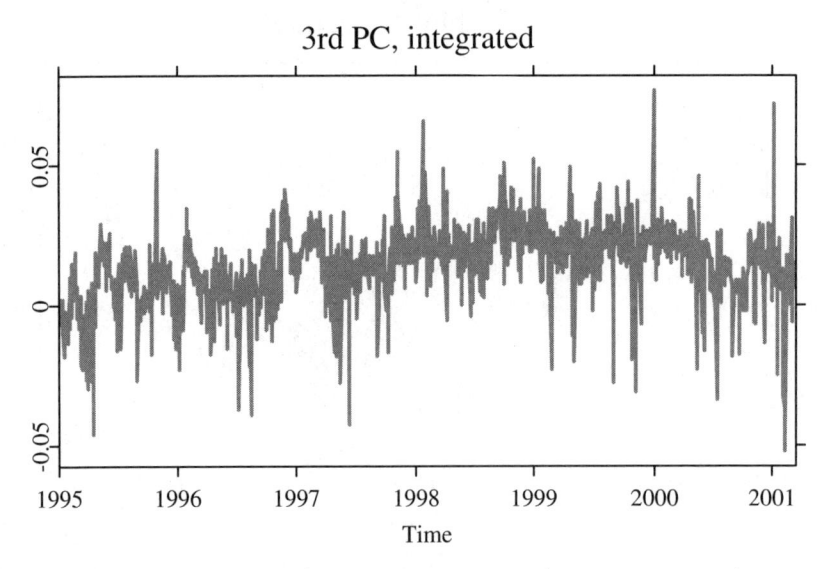

Fig. 5.6. Projection of the longest maturity group (90 days to expiry) using the third eigenvector. The *upper panel* shows returns, the *lower panel* the integrated series

by Zhu and Avellaneda (1997), Skiadopoulos et al. (1999), Alexander (2001b), Cont and da Fonseca (2002), Fengler et al. (2002b). The interpretations of the factor loadings in terms of shift, slope and twist shocks are also known from PCA studies on interest and forward rates, Bliss (1997) and Rebonato (1998).

Table 5.4. Descriptive statistics of the first three PCs, 90 days to expiry

PC	Variance Explained	Standard Deviation	Skewness	Kurtosis	Mean Reversion	Corr. With Index
1	0.88	0.078	0.34	4.12	227.7	−0.48
2	0.06	0.020	0.30	6.54	36.5	0.08
3	0.03	0.015	0.22	7.30	2.2	−0.03

Table 5.4 summarizes the descriptive statistics of the PCs. The results are similar to the findings of Cont and da Fonseca (2002) reported on the S&P 500 and the FTSE 100, except for the mean reversion. Whereas skewness is close to zero for the three PCs, there is evidence for excess kurtosis especially for the second and third PC. The mean reversion of the integrated first PC is found to be around 230 days, i.e. almost a year, while the second PC exhibits a more short-lived mean reversion of 36 days. The third PC has a mean reversion of 2 days. To our experience, however, the estimates of the mean reversion tend to be very sensitive to the sample size chosen, and change significantly in annual subsamples. Thus, the estimates of the mean reversion coefficient should be taken with caution.

The correlation with the returns of underlying is around −0.5 for the first PC. This is in line with the leverage effect: according to this argument (implied) volatility rises, when there is a negative shock in the market value of the firm, since this results in an increase in the debt-equity ratio. For the second and third PCs, the correlations are negligible.

Stability Analysis Among Different Samples

For any application in trading or risk management model stability is a decisive model characteristic, since otherwise model risk is unreckonable. In terms of the CPC models, there are two types of stability of interest: the more important one refers to the stability of the transformation matrices $\mathbf{\Gamma}$. Stability of $\mathbf{\Gamma}$ implies that the model can be estimated in a given (historical) sample period and contemporaneous PCs can be obtained by daily updating the data base of IVs and by projecting them into the same space without explicitly estimating $\mathbf{\Gamma}$ again. The second type of stability refers to the variances of the components collected in $\mathbf{\Lambda}_i$. Instability of $\mathbf{\Lambda}_i$ does not imply the need to often re-estimate the model, since the CPC model places no restrictions on $\mathbf{\Lambda}_i$: $\mathbf{\Psi}_i = \mathbf{\Gamma}\mathbf{\Lambda}_i\mathbf{\Gamma}^\top$ may very well hold across time in the sense of time-dependent variances $\mathbf{\Psi}_{i,t}$ and $\mathbf{\Lambda}_{i,t}$. However, instability of $\mathbf{\Lambda}_i$ implies the need of time series models capturing the heteroscedasticity of the PCs and thus has an impact on the choice of the time series model of the PCs.

To assess stability we split the entire sample into $R = 7$ annual, non-overlapping subsamples with around 250 observations in each subsample, except for the last one with 105 observations. In each of the annual samples, we estimate the CPC model separately.

146   5 Dimension-Reduced Modeling

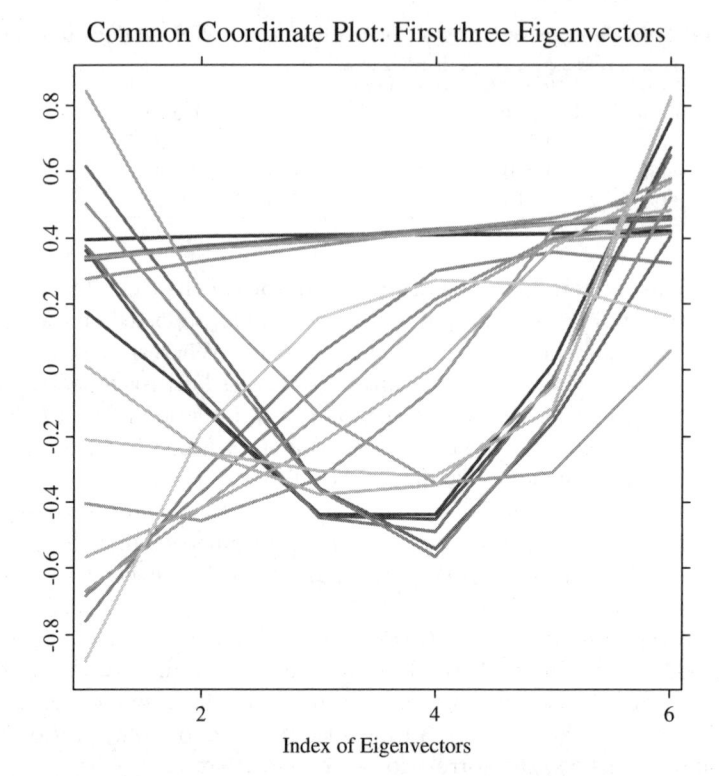

Common Coordinate Plot: First three Eigenvectors

Fig. 5.7. CPC model estimated separately in each annual sample 1995, 1996, 1997, 1998, 1999, 2000, 2001. Colors move from light to intensive tones the more recent the subsample

Let us first address stability among eigenvectors: in Fig. 5.7, we display the estimation results, where again blue refers to the first, green to the second and red to the third eigenvector. To highlight time dependence, colors move gradually from light to intensive tones the more recent the subsample.

As is immediately seen, the general structure of the eigenvectors is not altered: shift, slope and twist interpretation are visible for each sample. However, estimates display variability of different degrees. The first eigenvectors (blue) changes only little and appears to wander around the ATM IV thereby giving more equal weights to IVs across different moneyness for the most recent samples. The second and third eigenvector exhibit greater variance through time. For the years 1995, 1999, 2000, 2001 the second eigenvector appears concave, for 1996, 1997, 1998 convex. From the color intensity it is also seen that the third eigenvectors appear hardly altered for the most recent samples, whereas only those belonging to the samples 1995 and 1996 are largely different.

5.2 Common Principal Component Analysis 147

In a first testing attempt, we constructed a test for stability for all seven subsamples in one big test. However, it turned out that in this encompassing test, the variance-covariance matrix tended to be ill-conditioned, which caused numerical problems in computing its inverse. Accordingly, we decided to proceed sequentially in testing subsequently each year against the next following one. There are two caveats in this procedure: first, when there is a small but persistent trend in the data, it may be that the deviations from one year to the next one are too little to be detected by the test. However, over a longer horizon a large deviation may accumulate. To capture this possible effect, we choose the oldest subsample from 1995 as a benchmark, and test each eigenvector also against the 1995 estimates. As a second peril, we enter the well-known statistical problem of pre-tests. However, given the numerical difficulties encountered in the single test, we think that the sequential procedure is more reliable and can be justified. Furthermore, from a financial point of view, the information that significant changes occur from one year to the next may be of more interest than the information that something happened within the past seven years, since at some point – given the general concern of model risk – one will update or recalibrate the model in any case.

In Table 5.5, we present the test-statistics and the p-values of our tests. For the first eigenvector, against the benchmark year 1995, the stability hypothesis cannot be rejected at the 5% level of significance except for the years 2000 and 2001. The sequential tests reveal that there is a significant change from 1996 to 1997 and from 1999 to 2000. Our interpretation of these results, together with the visual inspection of Fig. 5.7, is that the first eigenvector is relatively reliable across the sample periods.

For the second and third eigenvectors, as can be conjectured from Fig. 5.7, the case is much different: the stability hypothesis is strongly rejected against the benchmark year. In the sequential tests, only from year 2000 to 2001 the null hypothesis cannot be rejected. There is a marginal case with respect to the third eigenvector, from 1997 to 1998. Altogether, we conclude that the second and third eigenvectors exhibit significant changes over time.

In the stability case of the eigenvalues, we present tests of only one group. There would be – if we tested three eigenvalues only – 132 tests (benchmark and sequential tests) to study in four groups with a moneyness grid of dimension six. Table 5.6 displays the results of the group with the shortest time to maturity (22 days to expiry). Results for the other groups are very similar. This is not surprising given the high degree of co-movements in the IVS. As is seen in Table 5.6, the null hypothesis is rejected against the benchmark year for all three eigenvalues. For the sequential tests, results are mixed: mostly all tests reject, but e.g. between the years of financial crisis 1997 and 1998, differences between the two samples are not significant in the first and second eigenvalue. As a general bottom line, for the second eigenvalue, differences between the years seem to be much less important than for the first and third one. This is an interesting result since it says – given our interpretation of this

Table 5.5. Stability tests of eigenvectors. Tests are constructed as derived in (5.25). The p-value is from a chi-squared variate with six degrees of freedom

First Eigenvectors

Sample 1	Sample 2	T	p-val.	Sample 1	Sample 2	T	p-val.
1995	1996	11.8	0.066				
1995	1997	5.6	0.465	1996	1997	28.8	0.000
1995	1998	12.5	0.051	1997	1998	2.4	0.873
1995	1999	6.6	0.352	1998	1999	4.7	0.580
1995	2000	29.2	0.000	1999	2000	39.1	0.000
1995	2001	22.6	0.001	2000	2001	1.38	0.966

Second eigenvectors

Sample 1	Sample 2	T	p-val.	Sample 1	Sample 2	T	p-val.
1995	1996	205.9	0.000				
1995	1997	188.8	0.000	1996	1997	289.0	0.000
1995	1998	85.8	0.000	1997	1998	19.6	0.003
1995	1999	539.1	0.000	1998	1999	99.3	0.000
1995	2000	100.8	0.000	1999	2000	173.8	0.000
1995	2001	28.4	0.000	2000	2001	9.34	0.155

Third eigenvectors

Sample 1	Sample 2	T	p-val.	Sample 1	Sample 2	T	p-val.
1995	1996	444.3	0.000				
1995	1997	108.0	0.000	1996	1997	532.1	0.000
1995	1998	36.4	0.000	1997	1998	16.7	0.010
1995	1999	251.8	0.000	1998	1999	66.7	0.000
1995	2000	48.1	0.000	1999	2000	92.9	0.000
1995	2001	19.8	0.000	2000	2001	7.4	0.284

component earlier – that volatility in the wings of the IVS is more constant than in the level and the twist component.

Summing up, from the stability tests, we draw the following conclusions: stability of the eigenvalues – except for the second one – is rejected. This is not a particular threat to modeling PCs or PCA in general, since it simply indicates that GARCH-type models can be an adequate choice in the time series context. For the eigenvectors, things look different: the good news is that the first eigenvector, the component, which captures more than 80% of the variance, is fairly stable. Thus in applications of risk controlling, such as scenario analysis or stress tests, see e.g. Fengler et al. (2002b), one can build on reliable estimates. The results from these experiments may not be completely correct in the wings of the IVS. However, since the biggest threat to option portfolios stems from level changes, the risk may be bearable from a risk management point of view. The bad news applies to trading strategies

5.2 Common Principal Component Analysis

Table 5.6. Stability tests of eigenvalues in group 1. Tests are constructed as derived in (5.19). The p-value is from a chi-squared variate with one degree of freedom

First Eigenvalues in Group 1

Sample 1	Sample 2	T	p-val.	Sample 1	Sample 2	T	p-val.
1995	1996	16.3	0.000				
1995	1997	41.6	0.000	1996	1997	10.3	0.001
1995	1998	54.3	0.000	1997	1998	2.5	0.108
1995	1999	33.8	0.000	1998	1999	6.8	0.009
1995	2000	15.0	0.000	1999	2000	6.1	0.013
1995	2001	18.7	0.000	2000	2001	5.8	0.016

Second eigenvalues in group 1

Sample 1	Sample 2	T	p-val.	Sample 1	Sample 2	T	p-val.
1995	1996	17.1	0.000				
1995	1997	18.4	0.000	1996	1997	52.4	0.000
1995	1998	55.1	0.000	1997	1998	19.9	0.108
1995	1999	56.9	0.000	1998	1999	0.1	0.807
1995	2000	62.5	0.000	1999	2000	0.6	0.418
1995	2001	71.0	0.000	2000	2001	2.5	0.109

Third eigenvalues in group 1

Sample 1	Sample 2	T	p-val.	Sample 1	Sample 2	T	p-val.
1995	1996	44.3	0.000				
1995	1997	35.7	0.000	1996	1997	87.6	0.000
1995	1998	72.5	0.000	1997	1998	22.4	0.000
1995	1999	40.7	0.000	1998	1999	18.0	0.000
1995	2000	79.1	0.000	1999	2000	26.8	0.000
1995	2001	83.6	0.000	2000	2001	1.3	0.240

that aim at exploiting the wings of the IVS, i.e. trading in OTM puts or OTM calls. Here, continuous recalibration of the models appears to be mandatory.

From our point of view, the results call for *adaptive techniques* of PCA that identify homogenous subintervals in the sample period by data-driven methods. On homogenous subintervals, reliable estimates are recovered. The literature on adaptive estimation, as pioneered by Lepski and Spokoiny (1997) and Spokoiny (1998), has been applied successfully in other contexts in finance such as time-inhomogenous volatility modeling, Härdle et al. (2003) and Mercurio and Spokoiny (2004).

Time Series Models

Due to the similarity within the groups (we consider the time series as scaled versions of each other), we concentrate on one group only. We pick the longest time to maturity group. The time series of the first three PCs $\mathbf{y}_{k1}, \mathbf{y}_{k3}, \mathbf{y}_{k3}$ are obtained from the projection $\mathbf{Y}_k = \mathbf{X}_k \hat{\mathbf{\Gamma}}$. Since the stability of the second and

ACF 1st PC

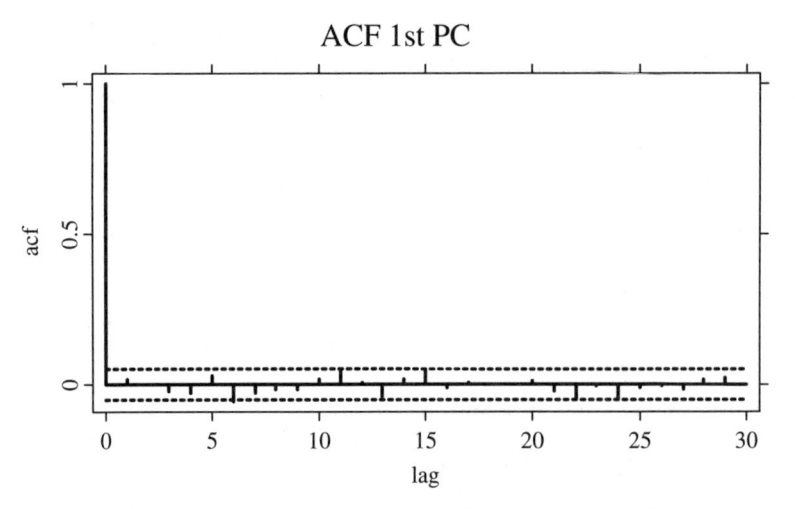

Fig. 5.8. Autocorrelation function of the first PC

third eigenvectors was rejected, we reestimate the model in each subsample and project using the new matrices $\hat{\mathbf{\Gamma}}^{(r)}$. Based on autocorrelation and partial autocorrelation plots, we propose adequate models for each univariate series. Of course the univariate time series are not independent, but they are uncorrelated by construction. This is why modeling the univariate series can be justified, see Zhu and Avellaneda (1997) for a similar approach. By AIC and SIC searches we will identify a best fitting model, and present the estimation results in more detail.

From Figs. 5.4 to 5.6, it is seen that the first three PCs display a behavior close to white noise. This impression is reinforced when inspecting the autocorrelation and partial autocorrelation functions as displayed in Figs. 5.8 to 5.13. From Fig. 5.8 it is seen that the first component exhibits no autocorrelation: it immediately dies off. Also the partial autocorrelation function in Fig. 5.9 does not show a particular structure. Thus, the first component, which explains up to 88% of the variance, can be considered as noise.

For the second and third components a different picture arises: from Figs. 5.10 and 5.12 a negative first order correlation is visible hinting towards an MA(1) model. Also the partial autocorrelation functions in Figs. 5.11 and 5.13 display the typical patterns of an MA process.

With this preliminary analysis at hand, we perform AIC and SIC searches over MA(q)-GARCH(r, s) models, where $q = 0$, $r = 1, 2$ $s = 1, 2$ for the first, and $q = 1$, $r = 1, 2$, $s = 1, 2$ for the second and third component. We also estimate different types of GARCH models such as TGARCH specifications in order to investigate asymmetries in shocks. Since Table 5.4 suggests a substantial correlation with the contemporaneous index returns, we additionally include index returns into the mean equations of all processes, and additionally into the variance equation of the first component.

5.2 Common Principal Component Analysis

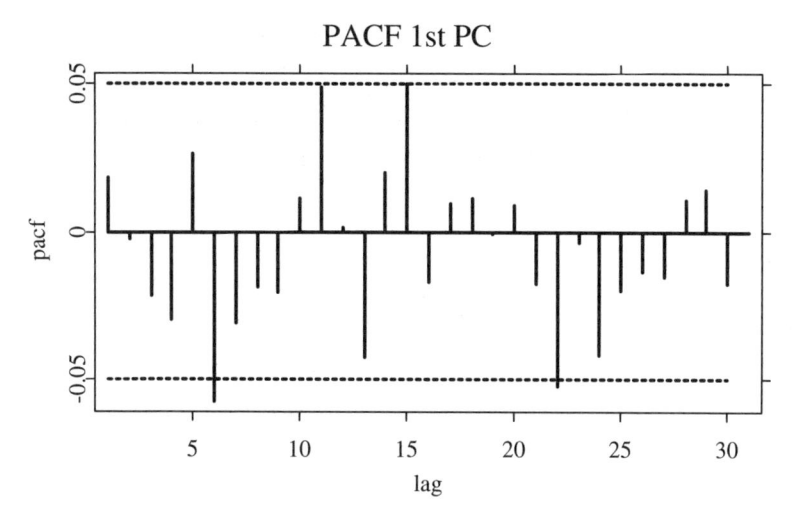

PACF 1st PC

Fig. 5.9. Partial autocorrelation function of the first PC

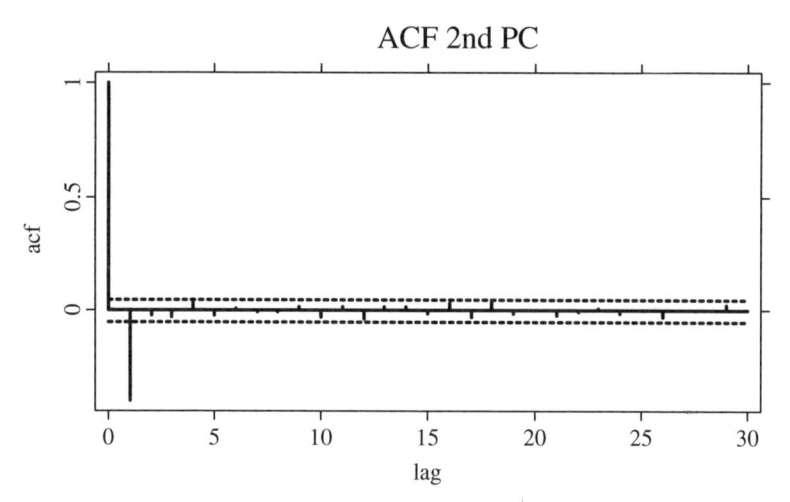

ACF 2nd PC

Fig. 5.10. Autocorrelation function of the second PC

The MA-GARCH models for the components $j = 1, 2, 3$ are given by:

$$y_{jt} = c + a_1 z_t + \varepsilon_{jt} + b_1 \varepsilon_{j,t-1};, \tag{5.31}$$
$$\varepsilon_{jt} \sim N(0, \sigma_{jt}^2),$$
$$\sigma_{jt}^2 = c_\sigma + \sum_{m=1}^{r} \alpha_m \sigma_{j,t-m} + \sum_{m=1}^{s} \beta_m \varepsilon_{j,t-m}^2 + \gamma z_t^2, \tag{5.32}$$

where we denote the elements of $\mathbf{y}_{k1}, \mathbf{y}_{k3}, \mathbf{y}_{k3}$ by y_{1t}, y_{2t}, y_{3t} to put ourselves into the usual time series notation. Log-returns in the DAX index are denoted by z_t.

PACF 2nd PC

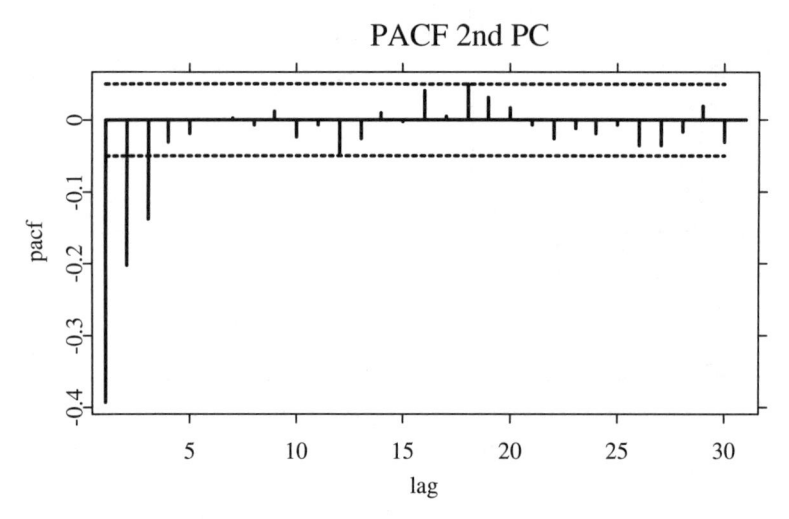

Fig. 5.11. Partial autocorrelation function of the second PC

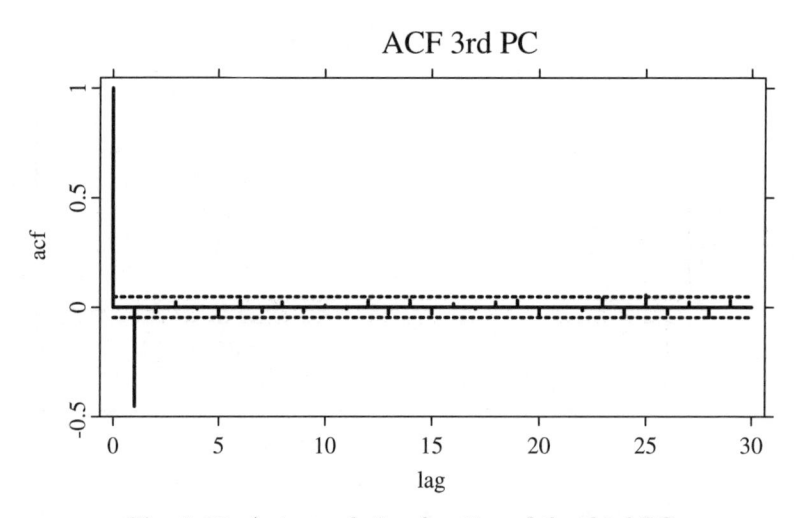

ACF 3rd PC

Fig. 5.12. Autocorrelation function of the third PC

Table 5.7 displays the statistics of the model selection criteria for the different models under consideration. For y_{1t} both AIC and SIC suggest an GARCH(1,2) specification. For y_{2t} and y_{3t}, the results are not as clear-cut. Since the differences of the model selection criteria are very much the same, we decided for the more parsimonious model, i.e. an MA(1)-GARCH(1,1) model for both.

Given these results, one may like to alter the variance equation to allow for asymmetries in shocks: under the TGARCH model, Glosten et al. (1993) and Zakoian (1994), the variance (5.32) becomes

5.2 Common Principal Component Analysis 153

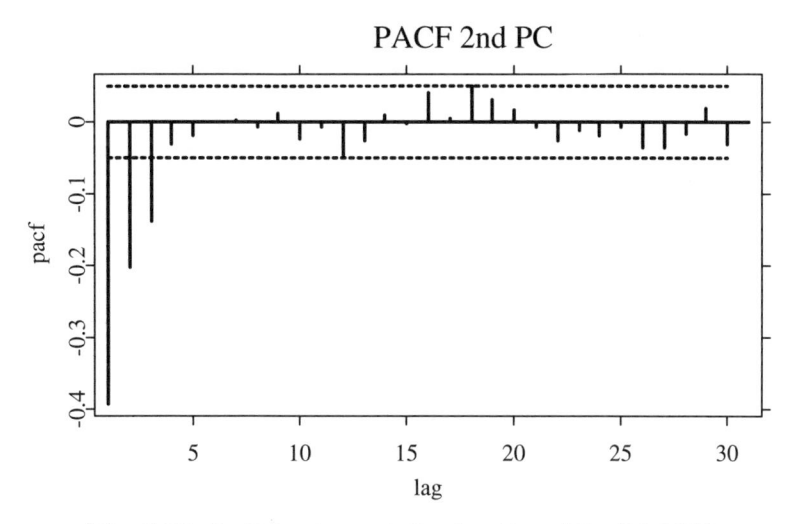

PACF 2nd PC

Fig. 5.13. Partial autocorrelation function of the third PC

Table 5.7. Univariate model selection: Akaike and Schwarz Information Criteria (AIC, SIC) over a variety of MA(q)-GARCH(r, s) models of y_{jt}

Model	AIC			SIC		
	y_{1t}	y_{2t}	y_{3t}	y_{1t}	y_{2t}	y_{3t}
GARCH(1,1)	−2.674			−2.654		
GARCH(1,2)	**−2.681**			**−2.657**		
GARCH(2,1)	−2.677			−2.654		
GARCH(2,2)	−2.681			−2.654		
MA(1)−GARCH(1,1)		−5.872	−6.460		−5.849	−6.436
MA(1)−GARCH(1,2)		−5.871	−6.460		−5.844	**−6.443**
MA(1)−GARCH(2,1)		−5.871	**−6.461**		−5.844	−6.434
MA(1)−GARCH(2,2)		−5.870	−6.461		−5.840	−6.431

$$\sigma_{jt}^2 = c_\sigma + \sum_{m=1}^{r} \alpha_m \sigma_{j,t-m} + \sum_{m=1}^{s} \beta_m \varepsilon_{j,t-m}^2 + \beta_1^- \varepsilon_{j,t-1}^2 \mathbf{1}(\varepsilon_{j,t-1} < 0) + z_t^2. \quad (5.33)$$

In this model, good news, $\varepsilon_t > 0$, and bad news, $\varepsilon_t < 0$, have differential effects on the conditional variance – good news have an impact of $\sum_{m=1}^{s} \beta_m$, while bad news have an impact of $\sum_{m=1}^{s} \beta_m + \beta_1^-$. If $\beta_1^- > 0$, a leverage effect exists, and the news impact is asymmetric if $\beta_1^- \neq 0$. We also estimated EGARCH models, Nelson (1991), however, since they did non produce any substantial gain compared to the other models, we do not report the estimation results here.

In Table 5.8, the estimation results are displayed in more detail. From the mean equation for y_{1t} it is evident that the index returns have a highly significant impact on the first PC. The sign is in line with the leverage effect hypothesis. In the variance equation all parameters are significant. $\beta_2 < 0$

Table 5.8. Estimation results of GARCH models for the three PCs, t-statistics in brackets

	Factor					
	y_{1t}		y_{2t}		y_{3t}	
cond. mean						
c	0.001	0.001	$1.9E^{-4}$	$1.0E^{-4}$	$-3.8E^{-05}$	$-5.8E^{-05}$
	[0.407]	[1.048]	[1.170]	[0.566]	[-0.592]	[-0.907]
a_1	-2.920	-2.930	0.086	0.079	0.005	0.004
	[-24.46]	[-24.21]	[4.860]	[4.564]	[0.457]	[0.351]
b_1			-0.733	-0.501	-0.733	-0.729
			[-35.50]	[-21.78]	[-35.50]	[-34.81]
cond. var.						
c_σ	$1.4E^{-4}$	$1.6E^{-4}$	$6.7E^{-5}$	$6.4E^{-5}$	$1.7E^{-05}$	$2.2E^{-05}$
	[3.945]	4.141	[7.515]	[7.353]	[8.687]	[8.681]
α_1	0.803	0.797	0.425	0.462	0.686	0.631
	[32.09]	[29.07]	[6.774]	[7.791]	[24.41]	[17.11]
β_1	0.246	0.284	0.200	0.115	0.147	0.082
	[7.112]	[7.598]	[6.840]	[3.505]	[8.027]	[3.206]
β_2	-0.130	-0.124				
	[-4.110]	[-3.611]				
β_1^-		-0.950		0.150		0.142
		[-3.706]		[3.239]		[3.916]
γ	1.480	1.580				
	[4.991]	[4.909]				
$\bar{R}^2$	0.23	0.23	0.22	0.21	0.33	0.33

may be interpreted as an 'over-reaction correction' in terms of the variance: high two-period lagged returns have a dampening impact on the variance. As is to be expected, volatility increases also when volatility in the underlying is high ($\gamma > 0$). From the TGARCH model, no evidence for a GARCH type leverage effect is found, since $\beta_1^- < 0$. The other parameter estimates for the TGARCH are of same size and significance level. The adjusted $\bar{R}^2$ is around 23%. This is high, however, it is entirely due to the index returns included in the regression. Leaving z_t out of the mean equations reduces the $\bar{R}^2$ to around 2%, only.

In the mean equations of y_{2t} and y_{3t}, the MA(1) components are negative and significant. The index returns are only significant for y_{2t} and positively influence the slope structure in the surface. Thus, together with the results for y_{1t}, we see that positive shocks in the underlying tend to reduce IV levels, while at the same time the slope of the surface is intensified. The variance equations do not exhibit any special features, however, it is interesting that a GARCH type leverage effect is present, since $\beta_1^- > 0$: lagged negative shocks increase the variance of both processes.

CPC Models: An Preliminary Summary

We have seen that CPC models yield a valid description of the IVS dynamics. They offer a convenient framework for model choice – and ultimately – for a low-dimensional description of the IVS. Three components that have intuitive financial interpretations as a shift, a slope and a twist shock appear to yield a sufficiently exact representation. Stability tests indicate that the first and most important component is fairly stable, while this conclusion cannot be drawn for the other two components. We employed GARCH models to describe the dynamics of the resulting factor series.

Within this framework risk and scenario analysis for portfolios can be implemented in a straightforward manner, Fengler et al. (2002b) and Fengler et al. (2003b). Forecasting is likely to be limited. At best a one-day forecast can be performed. Since this will be done in the context of the semiparametric factor model, we do not perform a separate forecast exercise at this point.

A potential disadvantage of CPC models is that the number of time series to be modelled are a multiple of the number of time to maturity groups, if one does not follow our simplification to model the series as scaled versions of each others. Also it would be more elegant, if factor extraction and surface estimation could be performed within a single step. This can be resolved by applying a functional PCA or using a semiparametric factor model as shall be seen presently.

5.3 Functional Data Analysis

Rethinking the approach taken in Sect. 5.2 suggests to carry the idea of PCA over to the functional case: this leads to functional PCA (FPCA). In PCA we obtain eigenvectors which are used to project the slices of the IVS into a lower dimensional space. In FPCA, we will recover eigenfunctions, or eigenmodes, for this projection (now defined in a functional sense). Similarly to PCA, we can represent the IVS as a linear combination of uncorrelated (scalar) random variables, which – via their eigenfunctions – unfold the high-dimensional dynamics of the IVS. In the literature of signal processing this representation is often called *Karhunen-Loève expansion* or *decomposition*. In the following subsection the basic ideas of the FPCA approach will be sketched. Key references for functional data analysis are Besse (1991) and Ramsay and Silverman (1997), who coined the field of functional data analysis. We also briefly address ways of computing FPCs. First application in the context of the IVS is due to Cont and da Fonseca (2002) who studied the IVS derived from options on the S&P 500 index and the FTSE 100 index. A treatment that also focusses on the computational aspects of FPCA, is given by Benko and Härdle (2004).

5.3.1 Basic Set-Up of FPCA

We consider the L^2 Hilbert space $\mathcal{H}(\mathcal{J})$ on a bounded interval $\mathcal{J} \subset \mathbb{R}^2$, where $\mathcal{J} = [\kappa_{\min}, \kappa_{\max}] \times [\tau_{\min}, \tau_{\max}]$ represents a region of moneyness and time to maturity. To model the IVS, we concentrate on sufficiently smooth elements of $\mathcal{H}$ that we interpret as *surfaces* over $\mathcal{J}$.

The inner product on $\mathcal{H}$ is given by

$$\langle f, g \rangle \stackrel{\text{def}}{=} \int_{\mathcal{J}} f(u) g(u) \, du \, , \quad \text{for } f, g \in \mathcal{H}(\mathcal{J}) \, , \tag{5.34}$$

and the norm $\|\cdot\|$

$$\|f\| \stackrel{\text{def}}{=} \left(\int_{\mathcal{J}} f(u)^2 \, du \right)^{1/2} . \tag{5.35}$$

The notation of the inner product can be distinguished from the covariation process of two stochastic processes $\langle \cdot, \cdot \rangle_t$ which is indexed by t.

We interpret a random surface X as a random function such that each realization $\omega \in \Omega$ gives a smooth surface $X(\omega, \cdot) : \mathcal{J} \to \mathbb{R}$. Without loss of generality we assume that X is mean zero. For the precise probabilistic set-up, we refer to Dauxois et al. (1982) and Pezzulli and Silverman (1993).

One can derive FPCA in the same step-wise manner as is typically done in PCA in standard textbook treatments: find linear combinations, i.e. weight functions $\gamma(u)$, such that the projection

$$Y_1 = \int_{\mathcal{J}} \gamma(u) \, X(u) \, du = \langle \gamma_1, X \rangle \tag{5.36}$$

has maximum variance subject to $\|\gamma_1\| = 1$. Continue, by finding another weight function γ_2 such that $Y_2 = \langle \gamma_2, X \rangle$ has maximum variance subject to $\|\gamma_2\| = 1$ and is orthogonal to γ_1 in the sense that $\langle \gamma_2, \gamma_1 \rangle = 0$, and so on.

This leads to the following constrained optimization problem:

$$\begin{aligned}
\max \ \mathsf{Var}\langle \gamma_j, X \rangle &= \max \ \mathsf{E} \int_{\mathcal{J}} \gamma_j(u) X(u) \int_{\mathcal{J}} \gamma_j(v) X(v) \, du \, dv \\
&= \max \ \int_{\mathcal{J}} \gamma_j(u) \int_{\mathcal{J}} C(u, v) \, \gamma_j(v) \, du \, dv \\
&= \max \ \langle \gamma_j, A \gamma_j \rangle
\end{aligned} \tag{5.37}$$

subject to $\|\gamma_j\|^2 = 1$ and $\langle \gamma_{j'}, \gamma_j \rangle = 0$ for $j' < j$. The covariance between the two surface values at $u, v \in \mathcal{J}$ is denoted by $C(u, v) \stackrel{\text{def}}{=} \mathsf{Cov}\{X(u), X(v)\}$, and the integral transform of the weight function γ with kernel C is defined by:

$$A\gamma(\cdot) \stackrel{\text{def}}{=} \int_{\mathcal{J}} C(\cdot, v) \, \gamma(v) \, dv \, . \tag{5.38}$$

We call the integral transform A which acts on γ the *covariance operator*. Since $C(\cdot,\cdot)$ is continuous and $\mathcal{J}$ bounded, A is compact. By definition, we have that A is symmetric and positive.

By general results from functional analysis, Riesz and Nagy (1956), the solution to this problem is obtained by solving the functional eigenvalue problem:

$$\int_{\mathcal{J}} C(u,v)\,\gamma_j(v)\,dv = \lambda_j\,\gamma_j(u)\,, \tag{5.39}$$

which is a Fredholm integral equation of the second kind. The sequence of eigenfunctions $\gamma_1, \gamma_2, \ldots$ and eigenvalues $\lambda_1 \geq \lambda_2 \geq \ldots \geq 0$ are the solutions to the maximization problem associated with FPCA. An important difference to multivariate PCA is the number of eigenfunction-eigenvalue pairs. In multivariate PCA, their number are equal to the number of variables measured: p in our former notation, whereas there are infinitely many in the functional case. In practice, the number depends on the rank of the covariance operator A.

The projection of X on $\gamma_j(u)$ is given by $Y_j = \langle \gamma_j, X \rangle$. By orthogonality of the sequence of eigenfunctions γ_j, the Y_j are a sequence of uncorrelated PCs. This implies that X is spanned by

$$X(u) = \sum_j Y_j\,\gamma_j(u)\,, \tag{5.40}$$

which yields the desired dimension reduction if the number of eigenfunctions, which are surfaces themselves, can be chosen to be small, see the discussion in Sect. 5.2.1, in particular (5.4), and Cont and da Fonseca (2002).

Finally, as in multivariate PCA, compare (5.3), the following link between the eigenvalues and the variance of the components holds:

$$\mathsf{Var}\langle \gamma_j, X\rangle = \langle \gamma_j, A\gamma_j \rangle = \lambda_j \|\gamma_j\|^2 = \lambda_j\,. \tag{5.41}$$

5.3.2 Computing FPCs

Denote by $x_i(u)$, $i = 1, 2, \ldots n$ and $u \in \mathcal{J}$ a sample of realizations of the IVS. As a first step, replace the unknown covariance function Cov by its sample analogue $\widehat{\mathsf{Cov}}$ in further maintaining the assumption of a zero mean:

$$\widehat{\mathsf{Cov}}\{X(u), X(v)\} \stackrel{\text{def}}{=} \frac{1}{n-1}\sum_{i=1}^{n} x_i(u)\,x_i(v)\,. \tag{5.42}$$

There are a number of methods for computing FPCs and solving (5.39), Ramsay and Silverman (1997). The first approach consists in discretizing the functions. In the simplest case when $\mathcal{J}$ is only a one-dimensional interval, say a particular smile or the ATM term structure, one can recover the values $x_i(u_1), x_i(u_2), \ldots, x_i(u_p)$ on a dense grid, and store the data in $(n \times p)$ matrix. Then an ordinary PCA is applied. Since in practice it can happen that $p >$

n, it may be necessary to recover the solution to the eigenvalue problem from the singular value decomposition of the data matrix. In order to recover the functional form of the eigenvectors, they are renormalized and suitably interpolated, Ramsay and Silverman (1997, Sect. 6.4.1). In principle, one could proceed similarly in the two-dimensional case where $\mathcal{J}$ contains the full region of moneyness by stacking the surfaces into a huge matrix. After applying an ordinary PCA, the resulting eigenvectors are resorted to recover the two-dimensional eigenfunctions.

Another, more elegant solution relies on basis expansions of the eigenfunctions, Ramsay and Silverman (1997, Section 6.4.2) and Cont and da Fonseca (2002). Suppose that the IVS admits an expansion in terms of a set of L basis functions $\phi_1(u), \phi_1(u), \ldots, \phi_L(u)$, $u \in \mathcal{J}$. Then each function is written as:

$$x_i(u) = \sum_{l=1}^{L} c_{il}\phi_l(u) , \qquad (5.43)$$

or, more compactly in matrix notation:

$$\mathbf{x}(u) = \mathbf{C}\boldsymbol{\phi}(u) , \qquad (5.44)$$

where the vectors $\mathbf{x}(u) \stackrel{\text{def}}{=} (x_i(u))$, and $\boldsymbol{\phi}(u) \stackrel{\text{def}}{=} (\phi_l(u))$, and the matrix $\mathbf{C} \stackrel{\text{def}}{=} (c_{il})$ for $i = 1, \ldots, n$ and $l = 1, \ldots, L$ are defined by their elements. In this case the covariance function is expressed as

$$\widehat{\mathrm{Cov}}\{X(u), X(v)\} \stackrel{\text{def}}{=} \frac{1}{n-1}\boldsymbol{\phi}^\top(u)\mathbf{C}^\top\mathbf{C}\boldsymbol{\phi}(v) . \qquad (5.45)$$

Similarly, the eigenfunction is expressed in terms of the basis functions as $\gamma(u) = \sum_{l=1}^{L} b_l\phi_l(u)$, or again in matrix form by $\gamma(u) = \boldsymbol{\phi}^\top(u)\mathbf{b}$.

With these preparations, one transforms the left-hand side of (5.39) to:

$$\frac{1}{n-1}\int_\mathcal{J} \boldsymbol{\phi}^\top(u)\,\mathbf{C}^\top\mathbf{C}\boldsymbol{\phi}(v)\,\boldsymbol{\phi}^\top(v)\,\mathbf{b}\,dv = \frac{1}{n-1}\boldsymbol{\phi}^\top(u)\,\mathbf{C}^\top\mathbf{C}\mathbf{W}\mathbf{b} , \qquad (5.46)$$

where $\mathbf{W} = (w_{l,l'}) \stackrel{\text{def}}{=} \int_\mathcal{J} \phi_l(v)\phi_{l'}(v)\,dv$. Thus, the (5.39) reads as

$$\frac{1}{n-1}\boldsymbol{\phi}^\top(u)\,\mathbf{C}^\top\mathbf{C}\mathbf{W}\mathbf{b} = \lambda\boldsymbol{\phi}^\top(u)\,\mathbf{b} . \qquad (5.47)$$

Since the last equation must hold for any $u \in \mathcal{J}$, it reduces to the pure matrix equation:

$$\frac{1}{n-1}\mathbf{C}^\top\mathbf{C}\mathbf{W}\mathbf{b} = \lambda\mathbf{b} . \qquad (5.48)$$

Equation (5.48) is further simplified by the following observation: in our basis framework, the inner product corresponds to

5.3 Functional Data Analysis

$$\langle \gamma_j, \gamma_{j'} \rangle = \int_{\mathcal{J}} \mathbf{b}_j^\top \boldsymbol{\phi}(u) \boldsymbol{\phi}^\top(u) \mathbf{b}_{j'} \, du = \mathbf{b}_j^\top \mathbf{W} \mathbf{b}_{j'} \,. \tag{5.49}$$

Defining $\mathbf{u} \stackrel{\text{def}}{=} \mathbf{W}^{1/2} \mathbf{b}$, one can transform (5.48) into the symmetric eigenvalue problem:

$$\frac{1}{n-1} \mathbf{W}^{1/2} \mathbf{C}^\top \mathbf{C} \mathbf{W}^{1/2} \mathbf{u} = \lambda \mathbf{u} \,. \tag{5.50}$$

This is solved using any standard PCA routines in statistical packages. The desired eigenfunctions are recovered by $\mathbf{b} = \mathbf{W}^{-1/2} \mathbf{u}$.

A special case occurs, when the basis functions are orthonormal. Then $\mathbf{W} = \mathbf{I}_L$, i.e. it becomes the identity matrix of order L. Hence, FPCA is reduced to the multivariate PCA performed on the coefficient matrix $\mathbf{C}$, Ramsay and Silverman (1997).

The concept of expanding the unknown solution to (5.39) on a set of basis functions is also known as *collocation*. It should be outlined that this superposition of basis functions leads to a strong solution of the underlying Fredholm integral equation of the second kind. This highlights the main difference to the well-known *Galerkin* methods that solve (5.39) in a weak sense, i.e. with respect to the corresponding dual space of $\mathcal{H}$. To implement the Galerkin method, one starts with a finite dimensional subspace of this dual space, and solves (5.39) with respect to a basis of this subspace. As the dimension of that subspace tends to infinity, one can obtain a solution that holds for all linear functionals. The Galerkin approach is taken by Cont and da Fonseca (2002).

To this end, Cont and da Fonseca (2002) expand the eigenfunctions up to an error on a basis

$$\gamma_j(u) = \sum_{l=1}^{L} b_{j,l} \phi_l(u) + \epsilon_j \,. \tag{5.51}$$

Plugging (5.51) into (5.39) yields, up to another error term:

$$\varepsilon_j = \sum_{l=1}^{L} b_{j,l} \left(\int_{\mathcal{J}} C(u,v) \phi_l(v) \, du - \lambda_j \phi_l(u) \right) \,. \tag{5.52}$$

It should be noted that implicitly both errors ϵ_j and ε_j depend on L, the number of basis functions.

The Galerkin approach requires the orthogonality of the error ε_j to the approximating functions ϕ_l, $l = 1, \ldots, L$, i.e.:

$$\langle \varepsilon_j, \phi_l \rangle = 0 \,. \tag{5.53}$$

This yields

$$\sum_{l=1}^{L} b_{j,l} \left(\int_{\mathcal{J}} \int_{\mathcal{J}} C(u,v) \phi_j(u) \phi_l(v) \, dv \, du - \lambda_j \int_{\mathcal{J}} \phi_j(u) \phi_l(u) \, du \right) = 0 \,. \tag{5.54}$$

Assuming that N eigenfunctions are to be recovered, introduce the following the matrix notation (in an elementwise sense):

$$\mathbf{B} = (b_{j,l}) \tag{5.55}$$

$$\mathbf{W} = (w_{j,l}) \stackrel{\text{def}}{=} \int_{\mathcal{J}} \phi_j(v)\,\phi_l(v)\,dv \tag{5.56}$$

$$\mathbf{C} = (c_{j,l}) \stackrel{\text{def}}{=} \int_{\mathcal{J}} \int_{\mathcal{J}} C(u,v) \phi_j(u) \phi_l(v)\,dv\,du \tag{5.57}$$

$$\mathbf{\Lambda} \stackrel{\text{def}}{=} \operatorname{diag}(\lambda_j,\, j = 1,\ldots,N)\,. \tag{5.58}$$

Then (5.54) can be summarized as

$$\mathbf{CB} = \mathbf{\Lambda W B}\,. \tag{5.59}$$

The solution of this generalized eigenvalue problem, $\mathbf{B}$ and $\mathbf{\Lambda}$, delivers the eigenfunctions by substituting into (5.51). The functional PCs are obtained via the projection (5.36), the associated variances of which are given by λ_j.

Cont and da Fonseca (2002) show that three eigenfunctions explain more than 95% of the variance of the IVS found in S&P 500 and FTSE 100 options. The particulars of their empirical evidence are very close to the results obtained from the DAX index options using either the CPC models or the semiparametric factor model, see Sect. 5.2 and Sect. 5.4, respectively.

5.4 Semiparametric Factor Models

In modeling the IVS one faces two main challenges: first, the data design is degenerated. Due to trading conventions, observations of the IVS occur only for a small number of maturities such as one, two, three, six, nine, twelve, 18, and 24 months to expiry on the date of issue. Consequently, IVs appear like pearls strung on a necklace – or in short – as strings. This pattern has been discussed in Sect. 2.5. For convenience, we display again the IVS together with a plot, which shows the data design as seen from the top, Fig. 5.14. Options belonging to the same string have a common time to maturity, i.e. lie on the same line. As time passes, the strings move through the maturity axis towards expiry, while changing levels and shape in a random fashion.

As a second challenge, also in the moneyness dimension, the observation grid does not cover the desired estimation grid at any point in time with the same density. Consider, for instance the third IV string from the bottom: only in a moneyness interval between 0.8 and 1.1 is occupied, while the coverage for the second string from the bottom is much wider. The reasons for this pattern can be twofold: first, these contracts have simply not been traded and consequently do not show up in a (transaction based) data set. The second reason – which is the more likely in this particular case – is hidden in the specific institutional arrangements at the futures exchange with regard to the

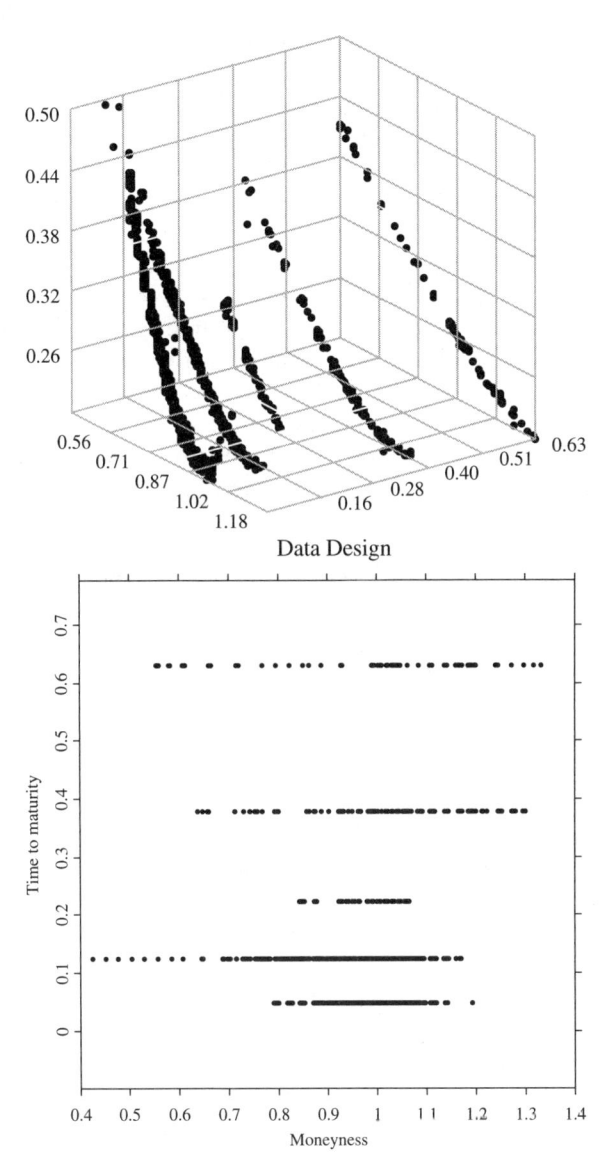

Fig. 5.14. *Top panel*: call and put IVs observed on 20000502. *Bottom panel*: data design on 20000502

creation of new contracts. Note that the options belonging to the third string expire in July and have been created at the beginning of April. When new contracts of a particular time to maturity are created, they are not available on the entire strike spectrum: initially, only a certain range of OTM and ITM options are open for trading. New contracts of this time to maturity are subsequently born, as the underlying price moves. This practice ensures that a minimum range of OTM and ITM options around the current spot price of the underlying asset is always maintained. In reference to Fig. 5.14, this means that contracts of other strikes may simply not exist, since the underlying moved too little between April and May.

Whatever the precise reasons are, it needs to be taken as a fact that even when the data sets are huge as ours, for a large number of cases IV observations are missing for certain subregions of the desired estimation grid. Of course, this is a point that will be most virulent in transaction based data sets.

The dimension reduction techniques from the previous sections fit the IVS on a grid for each day. Afterwards a PCA using a functional norm is applied to the surfaces. For the semi- or nonparametric approximations to the IVS, which are used within this work and which are promoted by Aït-Sahalia and Lo (1998), Rosenberg (2000), Aït-Sahalia et al. (2001b), Cont and da Fonseca (2002), Fengler et al. (2003b), and Fengler and Wang (2003), this design may pose difficulties. For illustration, consider in Fig. 5.15 (left panel) the fit of a standard Nadaraya-Watson estimator. Bandwidths are $h_1 = 0.03$ for the moneyness and $h_2 = 0.04$ for the time to maturity dimension (measured in years). The fit appears very rough, and there are huge holes in the surface, since the bandwidths are too small to 'bridge' the gaps between the maturity strings. In order to remedy this deficiency one would need to strongly increase the bandwidths. But this can induce a model bias. Moreover, since the design is time-varying, bandwidths would also need to be adjusted anew for each trading day, which complicates daily applications.

As an alternative, we will introduce the semiparametric factor model (SFM) with time-varying coefficients due to Fengler et al. (2003a). In this approach the IVS is fitted each day at the observed design points which will lead to a minimization with respect to functional norms that depend on time. This procedure avoids bias effects which can ensue from global daily fits used in standard FPCA. In the following, we present the model, discuss its estimation, and provide an empirical analysis for our data for the years 1998 to 2001.

5.4.1 The Model

We denote by $\mathcal{J} = [\kappa_{f\min}, \kappa_{f\max}] \times [\tau_{\min}, \tau_{\max}]$ a two-dimensional interval that represents a region of moneyness and time to maturity. Further, define (log)-IV as $y_{i,j} \stackrel{\text{def}}{=} \ln\{\widehat{\sigma}_{i,j}(\kappa_f, \tau)\}$, where for our transaction based volatility data set the index i is the number of the day ($i = 1, \ldots, I$), and $j = 1, \ldots, J_i$ is an intra-day numbering of the option traded on day i. The observations $y_{i,j}$

Model fit 20000502

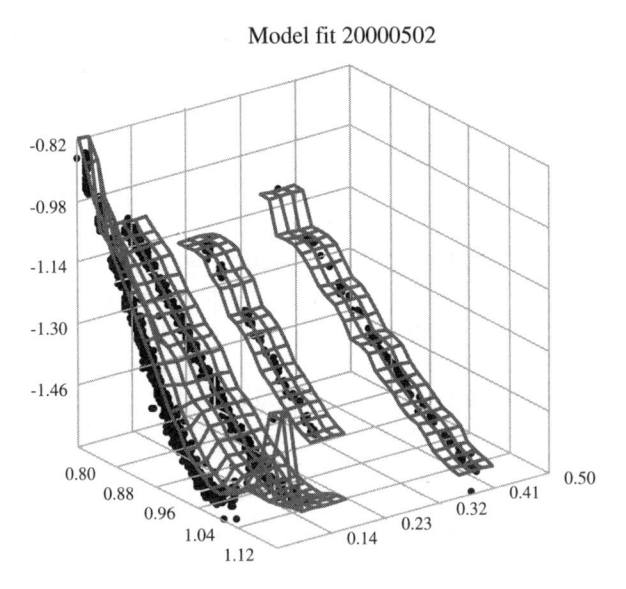

Semiparametric factor model fit 20000502

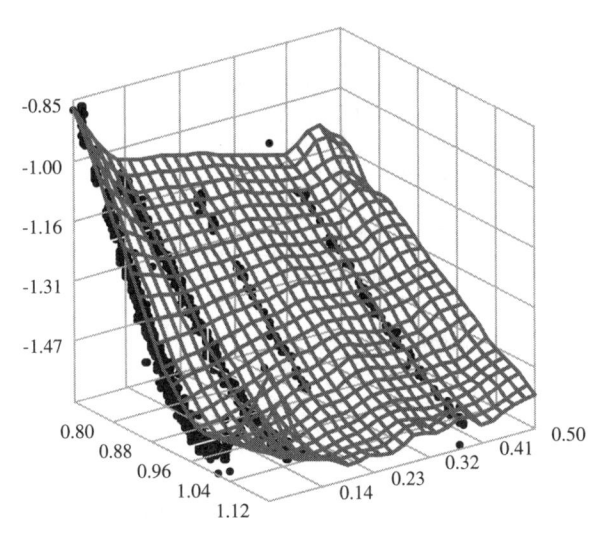

Fig. 5.15. Nadaraya-Watson estimate and SFM fit for 20000502. Bandwidths for both estimates $h_1 = 0.03$ for the moneyness and $h_2 = 0.04$ for the time to maturity dimension

are regressed on the two-dimensional covariables $\mathbf{x}_{i,j}$ that contain forward moneyness $\kappa_{f_{i,j}}$ and maturity $\tau_{i,j}$. The SFM approximates the IVS by:

$$y_{i,j} \approx m_0(\mathbf{x}_{i,j}) + \sum_{l=1}^{L} \beta_{i,l} m_l(\mathbf{x}_{i,j}) , \qquad (5.60)$$

where $m_l : \mathcal{J} \to \mathbb{R}$ are smooth basis functions ($l = 0, \ldots, L$). The IVS is approximated by a weighted sum of smooth functions m_l with weights $\beta_{i,l}$ depending on time i. The factor loading $\boldsymbol{\beta}_i \stackrel{\text{def}}{=} (\beta_{i,1}, \ldots \beta_{i,L})^\top$ forms an unobserved multivariate time series. By fitting the model (5.60), to the IV strings, we obtain approximations $\widehat{\boldsymbol{\beta}}_i$. We argue that VAR estimation based on $\widehat{\boldsymbol{\beta}}_i$ is asymptotically equivalent to an estimation based on the unobserved $\boldsymbol{\beta}_i$. After recovering the $\boldsymbol{\beta}_i$, we will model them in a suitable time series model. Hence, the time series of the factor loadings may be seen as *state variables*. This perspective reveals a close relationship of the model to Kalman filtering and is discussed in Borak et al. (2005).

In order to estimate the nonparametric components m_l and the state variables $\beta_{i,l}$ in (5.60), ideas from fitting additive models as in Stone (1986), Hastie and Tibshirani (1990), and Horowitz et al. (2002) are borrowed. The approach is related to functional coefficient models such as Cai et al. (2000). Other semi- and nonparametric factor models include Connor and Linton (2000), Gouriéroux and Jasiak (2001), Fan et al. (2003), and Linton et al. (2003) among others. Nonparametric techniques are now broadly used in option pricing, e.g. Broadie et al. (2000), Aït-Sahalia et al. (2001a), Aït-Sahalia and Duarte (2003), Daglish (2003), and interest rate modeling, e.g. Aït-Sahalia (1996), Ghysels and Ng (1989), and Linton et al. (2001).

Estimates $\widehat{m}_l$, ($l = 0, \ldots, L$) and $\widehat{\beta}_{i,l}$ ($i = 1, \ldots, I;\ l = 1, \ldots, L$) are defined as minimizers of the following least squares criterion ($\widehat{\beta}_{i,0} \stackrel{\text{def}}{=} 1$):

$$\sum_{i=1}^{I} \sum_{j=1}^{J_i} \int \left\{ y_{i,j} - \sum_{l=0}^{L} \widehat{\beta}_{i,l} \widehat{m}_l(u) \right\}^2 K_{\mathbf{h}}(u - \mathbf{x}_{i,j})\, du , \qquad (5.61)$$

where $u = (u_1, u_2) \in \mathcal{J}$. Further, $K_{\mathbf{h}}$ with $\mathbf{h} = (h_1, h_2)$ denotes the two-dimensional product kernel, $K_{\mathbf{h}}(u) \stackrel{\text{def}}{=} h_1^{-1} K_{(1)}(h_1^{-1} u_1) \times h_2^{-1} K_{(2)}(h_2^{-1} u_2)$, which is computed from one-dimensional kernels $K(v)$.

In (5.61) the minimization runs over all functions $\widehat{m}_l : \mathcal{J} \to \mathbb{R}$ and all values $\widehat{\beta}_{i,l} \in \mathbb{R}$. For illustration let us consider the case $L = 0$: the IVs $y_{i,j}$ are approximated by a surface $\widehat{m}_0$ that does not depend on time i. In this degenerated case, $\widehat{m}_0(u) = \sum_{i,j} K_{\mathbf{h}}(u - \mathbf{x}_{i,j}) y_{i,j} / \sum_{i,j} K_{\mathbf{h}}(u - \mathbf{x}_{i,j})$, which is the Nadaraya-Watson estimate based on the pooled sample of all days, compare with Sect. 4.2 and particularly with (4.13). In the algorithmic implementation of (5.61), the integral is replaced by Riemann sums on a fine grid.

Using (5.61) the IVS is approximated by surfaces moving in an L-dimensional affine function space $\{\widehat{m}_0 + \sum_{l=1}^{L} \alpha_l \widehat{m}_l : \alpha_1, \ldots, \alpha_L \in \mathbb{R}\}$. The

5.4 Semiparametric Factor Models

estimates $\widehat{m}_l$ are not uniquely defined: they can be replaced by functions that span the same affine space. In order to respond to this problem, we select $\widehat{m}_l$ such that they are orthogonal.

Replacing $\widehat{m}_l$ in (5.61) by $\widehat{m}_l + \delta g$ with arbitrary functions g and taking derivatives with respect to δ yields for $0 \leq l' \leq L$

$$\sum_{i=1}^{I} \sum_{j=1}^{J_i} \left\{ y_{i,j} - \sum_{l=0}^{L} \widehat{\beta}_{i,l} \widehat{m}_l(u) \right\} \widehat{\beta}_{i,l'} K_{\mathbf{h}}(u - \mathbf{x}_{i,j}) = 0 . \tag{5.62}$$

Furthermore, by replacing $\widehat{\beta}_{i,l}$ by $\widehat{\beta}_{i,l} + \delta$ in (5.61) and again taking derivatives with respect to δ, we get, for $1 \leq l' \leq L$ and $1 \leq i \leq I$:

$$\sum_{j=1}^{J_i} \int \left\{ y_{i,j} - \sum_{l=0}^{L} \widehat{\beta}_{i,l} \widehat{m}_l(u) \right\} \widehat{m}_{l'}(u) K_{\mathbf{h}}(u - \mathbf{x}_{i,j}) \, du = 0 . \tag{5.63}$$

Introducing the following notation for $1 \leq i \leq I$

$$\widehat{p}_i(u) = \frac{1}{J_i} \sum_{j=1}^{J_i} K_{\mathbf{h}}(u - \mathbf{x}_{i,j}) , \tag{5.64}$$

$$\widehat{q}_i(u) = \frac{1}{J_i} \sum_{j=1}^{J_i} K_{\mathbf{h}}(u - \mathbf{x}_{i,j}) y_{i,j} , \tag{5.65}$$

we obtain from (5.62)-(5.63) for $1 \leq l' \leq L, 1 \leq i \leq I$:

$$\sum_{i=1}^{I} J_i \widehat{\beta}_{i,l'} \widehat{q}_i(u) = \sum_{i=1}^{I} J_i \sum_{l=0}^{L} \widehat{\beta}_{i,l'} \widehat{\beta}_{i,l} \widehat{p}_i(u) \widehat{m}_l(u) , \tag{5.66}$$

$$\int \widehat{q}_i(u) \widehat{m}_{l'}(u) \, du = \sum_{l=0}^{L} \widehat{\beta}_{i,l} \int \widehat{p}_i(u) \widehat{m}_{l'}(u) \widehat{m}_l(u) \, du . \tag{5.67}$$

We calculate the estimates by iterative use of (5.66) and (5.67). We start with initial values $\widehat{\beta}_{i,l}^{(0)}$ for $\widehat{\beta}_{i,l}$. A possible choice of the initial $\widehat{\beta}_i$ could correspond to fits of an IVS that is piecewise constant on time intervals $I_1, \ldots, I_L$. This means, for $l = 1, \ldots, L$, put $\widehat{\beta}_{i,l}^{(0)} = 1$ (for $i \in I_l$), and $\widehat{\beta}_{i,l}^{(0)} = 0$ (for $i \notin I_l$). Here $I_1, \ldots, I_L$ are pairwise disjoint subsets of $\{1, \ldots, I\}$ and $\bigcup_{l=1}^{L} I_l$ is a strict subset of $\{1, \ldots, I\}$. For $r \geq 0$, we put $\widehat{\beta}_{i,0}^{(r)} = 1$. Define the matrix $\mathbf{B}^{(r)}(u)$ by its elements:

$$\left(b_{l,l'}^{(r)}(u) \right) \stackrel{\text{def}}{=} \sum_{i=1}^{I} J_i \widehat{\beta}_{i,l'}^{(r-1)} \widehat{\beta}_{i,l}^{(r-1)} \widehat{p}_i(u) , \quad 0 \leq l, l' \leq L , \tag{5.68}$$

166   5 Dimension-Reduced Modeling

and introduce a vector $\mathbf{q}^{(r)}(u)$ with elements

$$\left(q_l^{(r)}(u)\right) \stackrel{\text{def}}{=} \sum_{i=1}^{I} J_i \widehat{\beta}_{i,l}^{(r-1)} \widehat{q}_i(u) , \qquad 0 \leq l \leq L . \tag{5.69}$$

In the r-th iteration the estimate $\widehat{\mathbf{m}} = (\widehat{m}_0, \ldots, \widehat{m}_L)^\top$ is given by

$$\widehat{\mathbf{m}}^{(r)}(u) = \mathbf{B}^{(r)}(u)^{-1} \mathbf{q}^{(r)}(u) . \tag{5.70}$$

This update step is motivated by (5.66). The values of β are updated in the r-th cycle as follows: define the matrix $\mathbf{M}^{(r)}(i)$

$$\left(M_{l,l'}^{(r)}(i)\right) \stackrel{\text{def}}{=} \int \widehat{p}_i(u) \widehat{m}_{l'}^{(r)}(u) \widehat{m}_l^{(r)}(u) \, du , \qquad 1 \leq l, l' \leq L , \tag{5.71}$$

and define a vector $\mathbf{s}^{(r)}(i)$:

$$\left(s_l^{(r)}(i)\right) \stackrel{\text{def}}{=} \int \widehat{q}_i(u) \widehat{m}_l(u) \, du - \int \widehat{p}_i(u) \widehat{m}_0^{(r)}(u) \widehat{m}_l^{(r)}(u) \, du , \qquad 1 \leq l \leq L . \tag{5.72}$$

Motivated by (5.67), put

$$\left(\widehat{\beta}_{i,1}^{(r)}, \ldots, \widehat{\beta}_{i,L}^{(r)}\right)^\top = \mathbf{M}^{(r)}(i)^{-1} \mathbf{s}^{(r)}(i) . \tag{5.73}$$

The algorithm is run until only minor changes occur. In the implementation, we choose a grid of points and calculate $\widehat{m}_l$ at these points. In the calculation of $\mathbf{M}^{(r)}(i)$ and $\mathbf{s}^{(r)}(i)$, we replace the integral by a Riemann integral approximation using the values of the integrated functions at the grid points.

5.4.2 Norming of the Estimates

As discussed above, $\widehat{m}_l$ and $\widehat{\beta}_{i,l}$ are not uniquely defined. Therefore, we orthogonalize $\widehat{m}_0, \ldots, \widehat{m}_L$ in $L^2(\widehat{p})$, where $\widehat{p}(u) = I^{-1} \sum_{i=1}^{I} \widehat{p}_i(u)$, such that $\sum_{i=1}^{I} \widehat{\beta}_{i,1}^2$ is maximum, and given $\widehat{\beta}_{i,1}, \widehat{m}_0, \widehat{m}_1, \sum_{i=1}^{I} \widehat{\beta}_{i,2}^2$ is maximum, and so forth. These aims can be achieved by the following two steps: first replace

$$\widehat{m}_0 \text{ by } \widehat{m}_0^{\text{new}} = \widehat{m}_0 - \boldsymbol{\gamma}^\top \boldsymbol{\Gamma}^{-1} \widehat{\mathbf{m}} ,$$

$$\widehat{\mathbf{m}} \text{ by } \widehat{\mathbf{m}}^{\text{new}} = \boldsymbol{\Gamma}^{-1/2} \widehat{\mathbf{m}} , \tag{5.74}$$

$$\widehat{\boldsymbol{\beta}}_i \text{ by } \widehat{\boldsymbol{\beta}}_i^{\text{new}} = \boldsymbol{\Gamma}^{1/2} \left\{\widehat{\boldsymbol{\beta}}_i + \boldsymbol{\Gamma}^{-1} \boldsymbol{\gamma}\right\} ,$$

where we redefine the vector $\widehat{\mathbf{m}} = (\widehat{m}_1, \ldots, \widehat{m}_L)^\top$ not to contain $\widehat{m}_0$ any more. Further we define the $(L \times L)$ matrix $\boldsymbol{\Gamma} = \int \widehat{\mathbf{m}}(u) \widehat{\mathbf{m}}(u)^\top \widehat{p}(u) \, du$, or for clarity elementwise by $\boldsymbol{\Gamma} = (\gamma_{l,l'})$, with $\gamma_{l,l'} \stackrel{\text{def}}{=} \int \widehat{m}_l(u)\, \widehat{m}_{l'}(u)\hat{p}(u)du$. Finally, we have $\boldsymbol{\gamma} = (\gamma_l)$, with $\gamma_l \stackrel{\text{def}}{=} \int \widehat{m}_0(u)\widehat{m}_l(u)\hat{p}(u)\, du$.

Note that by applying (5.74), $\widehat{m}_0$ is replaced by a function that minimizes $\int \widehat{m}_0^2(u)\hat{p}(u)du$. This is evident because $\widehat{m}_0$ is orthogonal to the linear space spanned by $\widehat{m}_1, \ldots \widehat{m}_L$. By the second equation of (5.74), $\widehat{m}_1, \ldots, \widehat{m}_L$ are replaced by orthonormal functions in $L^2(\hat{p})$.

In a second step, we proceed as in PCA and define a matrix $\widetilde{\mathbf{B}}$ with elements $(\widetilde{b}_{l,l'}) = \sum_{i=1}^{I} \widehat{\beta}_{i,l}\widehat{\beta}_{i,l'}$ and calculate the eigenvalues of $\widetilde{\mathbf{B}}$, $\lambda_1 > \ldots > \lambda_L$, and the corresponding eigenvectors $\mathbf{z}_1, \ldots \mathbf{z}_L$. Put $\mathbf{Z} = (\mathbf{z}_1, \ldots, \mathbf{z}_L)$. Replace

$$\widehat{\mathbf{m}} \text{ by } \widehat{\mathbf{m}}^{\text{new}} = \mathbf{Z}^\top \widehat{\mathbf{m}}, \qquad (5.75)$$

(i.e. $\widehat{m}_l^{\text{new}} = \mathbf{z}_l^\top \widehat{\mathbf{m}}$), and

$$\widehat{\boldsymbol{\beta}}_i \text{ by } \widehat{\boldsymbol{\beta}}_i^{\text{new}} = \mathbf{Z}^\top \widehat{\boldsymbol{\beta}}_i. \qquad (5.76)$$

After the application of (5.75) and (5.76), the orthonormal basis of the model $\widehat{m}_1, \ldots, \widehat{m}_L$ is chosen such that $\sum_{i=1}^{I} \widehat{\beta}_{i,1}^2$ is maximum, and – given $\widehat{\beta}_{i,1}, \widehat{m}_0, \widehat{m}_1$ – the quantity $\sum_{i=1}^{I} \widehat{\beta}_{i,2}^2$ is maximum, and so on, i.e. $\widehat{m}_1$ is chosen such that as much as possible is explained by $\widehat{\beta}_{i,1}\widehat{m}_1$. Next $\widehat{m}_2$ is chosen to achieve the maximum explanation by $\widehat{\beta}_{i,1}\widehat{m}_1 + \widehat{\beta}_{i,2}\widehat{m}_2$, and so forth.

Unlike in Sect. 5.3.1 on FPCA, the functions $\widehat{m}_l$ are not eigenfunctions of an operator. This is because we use a different norm, namely $\int f^2(u)\hat{p}_i(u)du$, for each day. Through the norming procedure the functions are chosen as eigenfunctions in an L-dimensional approximating linear space. The L-dimensional approximating spaces are not necessarily nested for increasing L. For this reason the estimates cannot be calculated by an iterative procedure that starts by fitting a model with one component, and that uses the old $L-1$ components in the iteration step from $L-1$ to L to fit the next component. The calculation of $\widehat{m}_0, \ldots, \widehat{m}_L$ has to be redone for different choices of L.

5.4.3 Choice of Model Parameters

For the choice of L, we consider the residual sum of squares for different L:

$$RV(L) \stackrel{\text{def}}{=} \frac{\sum_i^I \sum_j^{J_i} \left\{ y_{i,j} - \sum_{l=0}^{L} \widehat{\beta}_{i,l} \widehat{m}_l(\mathbf{x}_{i,j}) \right\}^2}{\sum_i^I \sum_j^{J_i} (y_{i,j} - \bar{y})^2}, \qquad (5.77)$$

where $\bar{y}$ denotes the overall mean of the observations. The quantity $1 - RV(L)$ is the portion of variance explained in the approximation, and L can be increased until a sufficiently high level of fitting accuracy is achieved. As has been explained for the CPC models, see (5.30), this is a common selection method also in PCA.

168   5 Dimension-Reduced Modeling

For a data-driven choice of bandwidths, we propose an approach based on a weighted Akaike Information Criterion (AIC). We argue for using a weighted criterion, since the distribution of the observations is far from regular, as was seen from Fig. 5.16. As mentioned in Sect. 4.3, this leads to nonconvexity in the criterion and typically to inacceptably small bandwidths. Given the unequal distribution of observations, it is natural to punish the criterion in areas where the distribution is sparse. For a given weight function w, consider:

$$\triangle(m_0,\ldots,m_L) \stackrel{\text{def}}{=} \mathsf{E}\frac{1}{N}\sum_{i,j}\{y_{i,j} - \sum_{l=0}^{L}\beta_{i,l}m_l(\mathbf{x}_{i,j})\}^2 w(\mathbf{x}_{i,j}) , \qquad (5.78)$$

for functions $m_0,\ldots,m_L$. We choose bandwidths such that $\triangle(\widehat{m}_0,\ldots,\widehat{m}_L)$ is minimum. According to the AIC this is asymptotically equivalent to minimizing:

$$\Xi_{AIC_1} \stackrel{\text{def}}{=} \frac{1}{N}\sum_{i,j}\{y_{i,j} - \sum_{l=0}^{L}\widehat{\beta}_{i,l}\widehat{m}_l(\mathbf{x}_{i,j})\}^2 w(\mathbf{x}_{i,j})$$

$$\times \exp\left\{2\frac{L}{N}K_{\mathbf{h}}(0)\int w(u)du\right\} . \qquad (5.79)$$

Alternatively, one may consider the computationally easier criterion:

$$\Xi_{AIC_2} \stackrel{\text{def}}{=} \frac{1}{N}\sum_{i,j}\{y_{i,j} - \sum_{l=0}^{L}\widehat{\beta}_{i,l}\widehat{m}_l(\mathbf{x}_{i,j})\}^2$$

$$\times \exp\left\{2\frac{L}{N}K_{\mathbf{h}}(0)\frac{\int w(u)\,du}{\int w(u)p(u)\,du}\right\} . \qquad (5.80)$$

Putting $w(u) \stackrel{\text{def}}{=} 1$, delivers the common AIC, see in particular Sect. 4.4.1. This, however, does not take into account the quality of the estimation at the boundary regions or in regions where the data are sparse, since in these regions $p(u)$ is small. We propose to choose

$$w(u) \stackrel{\text{def}}{=} \frac{1}{p(u)} , \qquad (5.81)$$

which gives equal weight everywhere as can be seen by the following considerations:

5.4 Semiparametric Factor Models   169

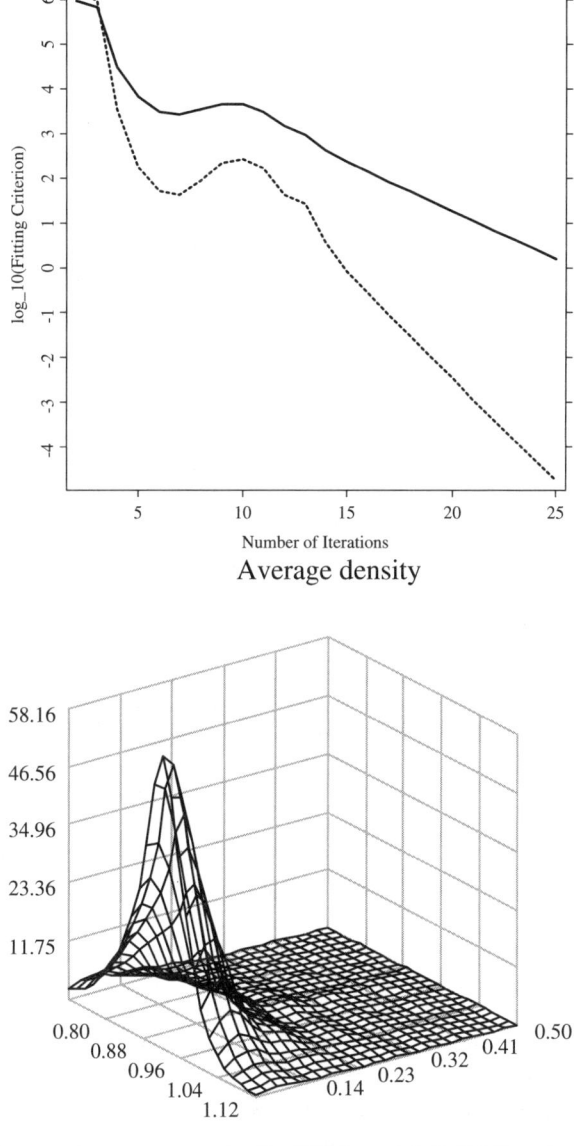

Fig. 5.16. *Top panel*: convergence in the SFM model. Solid line shows the L^1, the *dotted line* the L^2 measure of convergence. The total number of iterations are 25. *Bottom panel*: average density $\hat{p}(u) = I^{-1}\sum_{i=1}^{I}\hat{p}_i(u)$. Bandwidths are $h_1 = 0.03$ for moneyness and $h_2 = 0.04$ for time to maturity

$$\triangle(m_0,\ldots,m_L) = \mathsf{E}\frac{1}{N}\sum_{i,j}\varepsilon^2\, w(\mathbf{x}_{i,j})$$

$$+ \mathsf{E}\frac{1}{N}\sum_{i,j}\left[\sum_{l=0}^{L}\beta_{i,l}\{m_l(\mathbf{x}_{i,j}) - \widehat{m}_l(\mathbf{x}_{i,j})\}\right]^2 w(\mathbf{x}_{i,j})$$

$$\approx \sigma^2\int w(u)p(u)\,du$$

$$+ \frac{1}{N}\sum_{i,j}\int\left[\sum_{l=0}^{L}\beta_{i,l}\{m_l(u) - \widehat{m}_l(u)\}\right]^2 w(u)p(u)\,du\,. \tag{5.82}$$

The two criteria become:

$$\Xi_{AIC_1} \stackrel{\text{def}}{=} \frac{1}{N}\sum_{i,j}\left\{y_{i,j} - \sum_{l=0}^{L}\widehat{\beta}_{i,l}\widehat{m}_l(\mathbf{x}_{i,j})\right\}^2 \hat{p}(\mathbf{x}_{i,j})$$

$$\times\ \exp\left\{2\frac{L}{N}K_{\mathbf{h}}(0)\int\frac{1}{\hat{p}(u)}du\right\},\tag{5.83}$$

and

$$\Xi_{AIC_2} \stackrel{\text{def}}{=} \frac{1}{N}\sum_{i,j}\{y_{i,j} - \sum_{l=0}^{L}\widehat{\beta}_{i,l}\widehat{m}_l(\mathbf{x}_{i,j})\}^2$$

$$\times\ \exp\left\{2\frac{L}{N}K_{\mathbf{h}}(0)\mu_\lambda^{-1}\int\frac{1}{\hat{p}(u)}du\right\},\tag{5.84}$$

where $\mu_\lambda \stackrel{\text{def}}{=} (\kappa_{f\max} - \kappa_{f\min})(\tau_{\max} - \tau_{\min})$ denotes the Lebesgue measure of the design set $\mathcal{J}$.

Under some regularity conditions, the AIC is an asymptotically unbiased estimate of the mean average squared error (MASE), Sect. 4.4. In our setting it would be consistent if the density of $\mathbf{x}_{i,j}$ did not depend on day i. Due to the irregular design, this is an unrealistic assumption. For this reason, Ξ_{AIC_1} and Ξ_{AIC_2} estimate weighted versions of the MASE.

In our AIC, the penalty term does not punish for the number of parameters $\widehat{\beta}_{i,l}$ that are employed to model the time series. This can be neglected because we will use a finite dimensional model for the dynamics of $\beta_{i,l}$. The corresponding penalty term is negligible compared to the smoothing penalty term. A corrected penalty term that takes care of the parametric model of $\widehat{\beta}_{i,j}$ will be considered in the empirical part in Sect. 5.4.4 where the prediction performance is assessed.

Clearly the choice of $\mathbf{h}$ and L are not independent. From this point of view, one may think about minimizing (5.83) or (5.84) over both parameters. However, our practical experience shows that for a given L, changes in the criteria from a variation in **h** are small compared to a variation in L for a given **h**. To reduce the computational burden, we use (5.77) to determine the model size L, and then (5.83) and (5.84) to optimize **h** for a given L.

The convergence of the iteration cycles is measured by

$$Q_k(r) \stackrel{\text{def}}{=} \sum_{i=1}^{I} \int \left| \sum_{l=0}^{L} \widehat{\beta}_i^{(r)} \widehat{m}_l^{(r)}(u) - \widehat{\beta}_i^{(r-1)} \widehat{m}_l^{(r-1)}(u) \right|^k du \, . \tag{5.85}$$

As above (r) denotes the result from the rth cycle of the estimation. Here, we approximate the integral by a simple sum over the estimation grid. Putting $k = 1, 2$, we have an L^1 and an L^2 measure of convergence. Iterations are stopped when $Q_k(r) \leq \epsilon_k$ for some small $\epsilon > 0$.

5.4.4 Empirical Analysis

IVs are observed only for particular strings, but in practice, one thinks about them as being the observed values of an entire surface, the IVS. This is evident, when one likes to price and hedge over-the-counter options expiring at intermediate maturities. We model log-IV on $\mathbf{x}_{i,j} = (\kappa_{i,j}, \tau_{i,j})^\top$. Our estimation set $\mathcal{J}$ covers in moneyness $\kappa_f \in [0.80, 1.20]$ and in time to maturity $\tau \in [0.05, 0.5]$ measured in years.

In this model, we employ $L = 3$ basis functions, which capture around 96.0% of the variations in the IVS. We believe this to be of sufficiently high accuracy. Bandwidths used are $h_1 = 0.03$ for moneyness and $h_2 = 0.04$ for time to maturity. This choice is justified by Table 5.9 which presents estimates for the two AIC criteria. Both criterion functions become very flat near the minimum, especially Ξ_{AIC_1}. However, Ξ_{AIC_2} assumes its global minimum in the neighborhood of $\mathbf{h}^* = (0.03, 0.04)^\top$, which is why we opt for this pair of bandwidths. In Table 5.9, we also display a measure of how the factor loadings and the basis functions change relative to the optimal bandwidth $\mathbf{h}^*$. More precisely, we compute:

$$V_{\widehat{\beta}}(\mathbf{h}_k) = \sqrt{\sum_{l=0}^{L} \text{Var}\{|\widehat{\beta}_{i,l}(\mathbf{h}_k) - \widehat{\beta}_{i,l}(\mathbf{h}^*)|\}} \, , \tag{5.86}$$

$$\text{and} \quad V_{\widehat{m}}(\mathbf{h}_k) = \sqrt{\sum_{l=0}^{L} \text{Var}\{|\widehat{m}_l(u; \mathbf{h}_k) - \widehat{m}_l(u; \mathbf{h}^*)|\}} \, , \tag{5.87}$$

where $\mathbf{h}_k$ runs over the values given in Table 5.9, and $\text{Var}(x)$ denotes the variance of x. It is seen that changes in $\widehat{m}$ are 10 to 100 times higher in magnitude than those for $\widehat{\beta}$. This corroborates the approximation in (5.82) that treats the factor loadings as known.

In being able to choose such small bandwidths, the strength of the modeling approach is demonstrated: the bandwidth in the time to maturity dimension is so small that in the fit of a particular day, data from contracts

5 Dimension-Reduced Modeling

Table 5.9. Bandwidth selection via AIC as given in (5.83) and (5.84) for different choices of $\mathbf{h} = (h_1, h_2)^\top$: h_1 refers to moneyness and h_2 to time to maturity measured in years; the bandwidths chosen are highlighted in bold. In all cases $L = 3$. $V_{\widehat{\beta}}$ and $V_{\widehat{m}}$ measure the change in $\widehat{\beta}$ and $\widehat{m}$ as functions of $\mathbf{h}$ relative to the optimal bandwidth $\mathbf{h}^* = (0.03, 0.04)^\top$, compare (5.86) and (5.87)

h_1	h_2	Ξ_{AIC_1}	Ξ_{AIC_2}	$V_{\widehat{\beta}}$	$V_{\widehat{m}}$
0.01	0.02	0.000737	0.00151	0.015	0.938
0.01	0.04	0.000741	0.00150	0.003	0.579
0.01	0.06	0.000739	0.00152	0.005	0.416
0.01	0.08	0.000736	0.00163	0.011	0.434
0.02	0.02	0.001895	0.00237	0.104	3.098
0.02	0.04	0.000738	0.00150	0.001	0.181
0.02	0.06	0.000741	0.00151	0.004	0.196
0.02	0.08	0.000742	0.00156	0.008	0.279
0.02	0.10	0.000744	0.00162	0.011	0.339
0.03	0.02	0.002139	0.00256	0.111	3.050
0.03	**0.04**	**0.000739**	**0.00149**	–	–
0.03	0.06	0.000743	0.00152	0.004	0.180
0.03	0.08	0.000743	0.00156	0.008	0.273
0.03	0.10	0.000744	0.00162	0.011	0.337
0.04	0.02	0.002955	0.00323	0.138	3.017
0.04	0.04	0.000743	0.00151	0.001	0.088
0.04	0.06	0.000746	0.00154	0.005	0.211
0.04	0.08	0.000745	0.00157	0.008	0.293
0.04	0.10	0.000746	0.00163	0.012	0.353
0.05	0.02	0.003117	0.00341	0.142	2.962
0.05	0.04	0.000748	0.00155	0.001	0.148
0.05	0.06	0.000749	0.00157	0.005	0.241
0.05	0.08	0.000748	0.00160	0.008	0.312
0.05	0.10	0.000749	0.00167	0.012	0.368
0.06	0.02	0.003054	0.00343	0.139	2.923
0.06	0.04	0.000755	0.00160	0.002	0.193
0.06	0.06	0.000756	0.00163	0.005	0.268
0.06	0.08	0.000754	0.00166	0.009	0.330
0.06	0.10	0.000754	0.00172	0.012	0.383

with two adjacent time to maturities do not enter together $\widehat{p}_i(u)$ in (5.64) and $\widehat{q}_i(u)$ in (5.65). In fact, for a given u', the quantities $\widehat{p}_i(u')$ and $\widehat{q}_i(u')$ are zero most of the time, and only assume positive values for dates i when the observations are in the local neighborhood of u'. The same applies to the moneyness dimension. Of course, during the entire observation period I, it is mandatory that at least some observations for each u at some dates i are made.

In Fig. 5.16, we display the L^1 and L^2 measures of convergence. Convergence is achieved quickly. The iterations were stopped after 25 cycles, when the L^2 was less than 10^{-5}. Figures 5.17 to 5.19 display the functions $\widehat{m}_1$ to

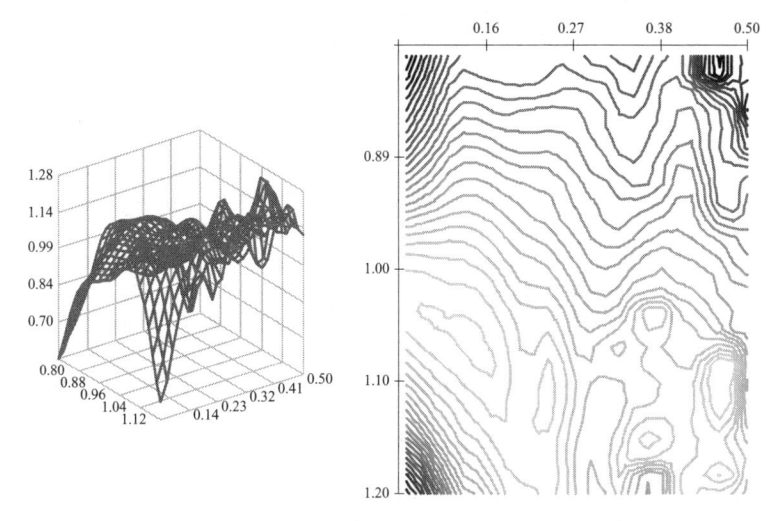

Fig. 5.17. Factor $\widehat{m}_1$ in the left panel (moneyness *lower left axis*). Right panel shows contour plots of this function (moneyness *left axis*). Lines are thick for positive level values, thin for negative ones. The gray scale becomes increasingly lighter the higher the level in absolute value. Stepwidth between contour lines is 0.028, estimated from ODAX data 19980101-20010531

$\widehat{m}_4$ together with contour plots. We do not display the invariant function $\widehat{m}_0$, since it essentially is the zero function of the affine space fitted by the data: both mean and median are zero up to 10^{-2} in magnitude. We believe this to be pure estimation error. The remaining functions exhibit more interesting patterns: $\widehat{m}_1$ in Fig. 5.17 is positive throughout, and mildly concave. There is little variability across the term structure. Since this function belongs to the weights with highest variance, we interpret it as the time dependent mean of the (log)-IVS, i.e. a *shift effect*. Clearly, these observations are (and must be) an iteration of the results from our CPC analysis in Sect. 5.2.5, see also Cont and da Fonseca (2002).

Function $\widehat{m}_2$, depicted in Fig. 5.18, changes sign around the ATM region, which implies that the smile deformation of the IVS is exacerbated or mitigated by this eigenfunction. Hence we consider this function as a *moneyness slope effect* of the IVS. Finally, $\widehat{m}_3$ is positive for the very short term contracts, and negative for contracts with maturity longer than 0.1 years, Fig. 5.19. Thus, a positive weight in $\widehat{\beta}_{i,3}$ lowers short term IVs and increases long term IVs: $\widehat{m}_3$ generates the term structure dynamics of the IVS, i.e. it provides a *term structure slope effect*.

To appreciate the power of the SFM, we inspect again the situation of 20000502. In Fig. 5.20 we compare a Nadaraya-Watson estimator (left panel) with the SFM (right panel). In the first case, the bandwidths are increased to $\mathbf{h} = (0.06, 0.25)^\top$ in order to remove all holes and excessive variation in the fit, while for the latter the bandwidths are kept at $\mathbf{h} = (0.03, 0.04)^\top$. While

174   5 Dimension-Reduced Modeling

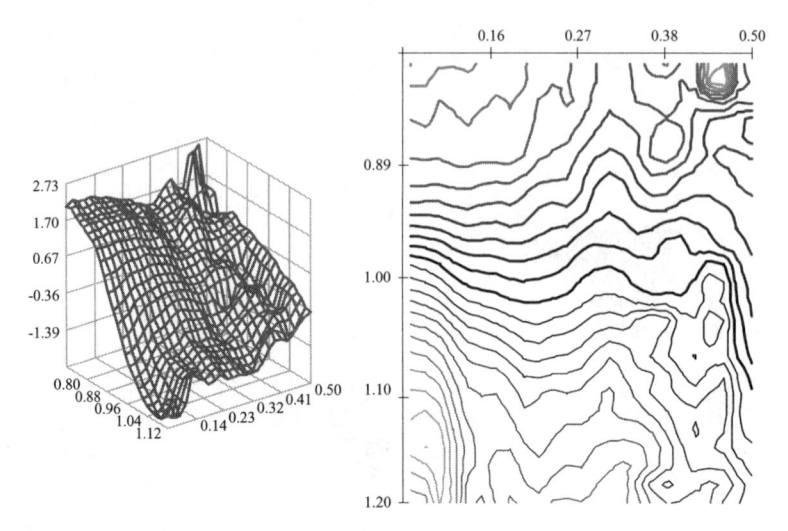

Fig. 5.18. Factor $\widehat{m}_2$ in the left panel (moneyness *lower left axis*). Right panel shows contour plots of this function (moneyness *left axis*). Lines are thick for positive level values, thin for negative ones. The gray scale becomes increasingly lighter the higher the level in absolute value. Stepwidth between contour lines is 0.225, estimated from ODAX data 19980101-20010531

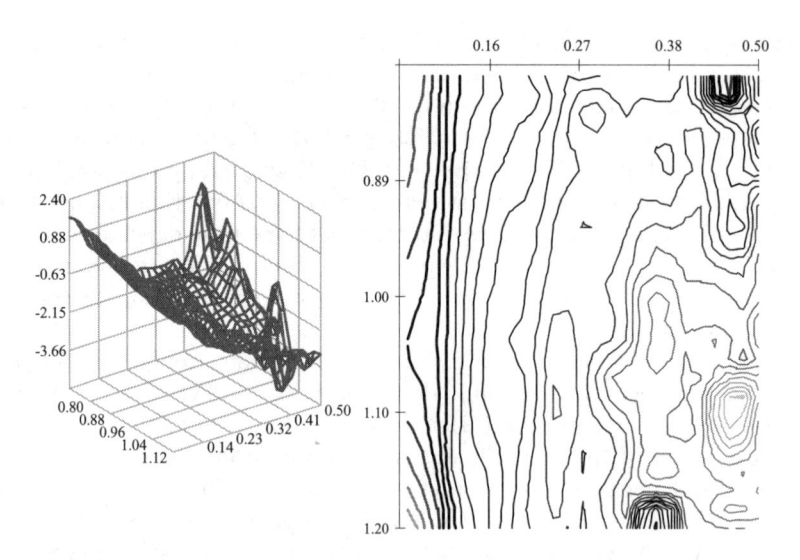

Fig. 5.19. Factor $\widehat{m}_3$ in the left panel (moneyness *lower left axis*). Right panel shows contour plots of this function (moneyness *left axis*). Lines are thick for positive level values, thin for negative ones. The gray scale becomes increasingly lighter the higher the level in absolute value. Stepwidth between contour lines is 0.240, estimated from ODAX data 19980101-20010531.

5.4 Semiparametric Factor Models 175

Model fit 20000502

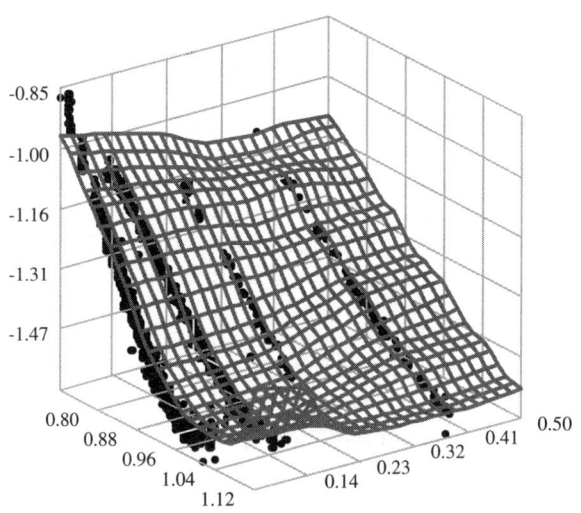

Semiparametric factor model fit 20000502

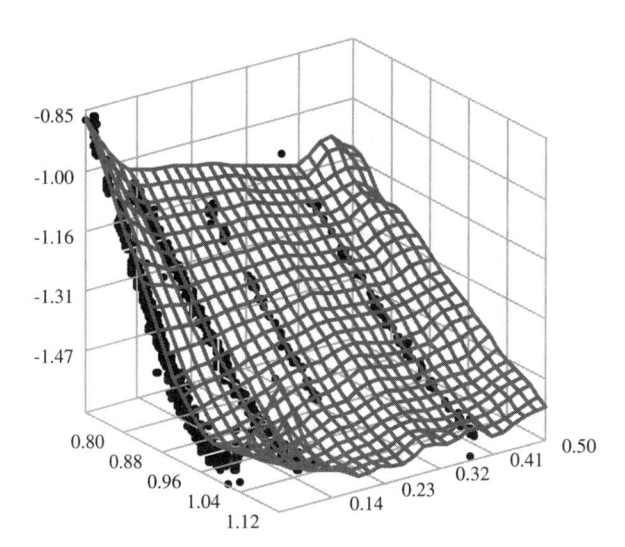

Fig. 5.20. Nadaraya-Watson estimator with $\mathbf{h} = (0.06, 0.25)^\top$ and SFM with $\mathbf{h} = (0.03, 0.04)^\top$ for 20000502

both fits look quite similar at a first glance, the differences are best visible when both cases are contrasted for each time to maturity string separately, Figs. 5.21 to 5.24. Note that these figures do *not* display separate fits of the smile functions. What we display are *slices* from the two-dimensional surfaces.

As is well seen, the standard Nadaraya-Watson fit exhibits a strong directional bias, especially in the wings of the IVS. For instance, for the short maturity contracts, Fig. 5.21, the estimated IVS is too low both in the OTM put and the OTM call region. At the same time, levels are too high for the 45 days to expiry contracts, Fig. 5.22. For the 80 days to expiry case, Fig. 5.23, the fit exhibits an S-formed shape, although the data lie almost on a linear line. Also the SFM is not entirely free of a directional bias, but clearly the fit is superior.

Figure 5.25 shows the entire time series of $\widehat{\beta}_{i,1}$ to $\widehat{\beta}_{i,3}$, the summary statistics are given in Table 5.10 and contemporaneous correlation in Table 5.11. The correlograms given in the lower panel of Fig. 5.25 display the rich autoregressive dynamics of the factor loadings. The ADF tests, Table 5.12, indicate a unit root for $\widehat{\beta}_{i,1}$ and $\widehat{\beta}_{i,2}$ at the 5% level. In following the pathway taken in Sect. 5.2.5 for the CPC models, one may model the first differences of the first two loading series together with the levels of $\widehat{\beta}_{i,3}$ in a parsimonious VAR framework. Alternatively, since the results are only marginally significant, one may estimate the levels of the loading series in a rich VAR model. Although our results from Sect. 5.2.5 also suggest a GARCH specification, we opt for the VAR model in levels. The main reason is that the loading series of the SFM – unlike those obtained from the CPC models – are not uncorrelated. Accordingly, one would need to specify a multivariate GARCH model. However, even for moderate dimensions the likelihood function of the multivariate GARCH model is quickly untractable or can deliver unstable results, Fengler and Herwartz (2002). As an alternative, one may consider dynamical correlation models. Introduced by Engle (2002) and Tse and Tsui (2002), they enjoy increasing popularity due to their tractability and richness of volatility and correlation patterns they allow for. We shall not pursue this model class at this point, but it may be profitable to do so in the future.

Given the preceding considerations we model the levels of the factor loadings in a VAR(2) model. The results are presented in Table 5.13. The estimation also includes a constant and two dummy variables, assuming the value one right at those days and one day after, when the corresponding IV observations of the minimum time to maturity string (10 days to expiry) were to be excluded from the estimation of the SFM, as is described in the Appendix A. This is to capture possible seasonality effects introduced from the data filter.

Estimation results are displayed in Table 5.13. In the equations of $\widehat{\beta}_{i,1}$ and $\widehat{\beta}_{i,2}$ the constants and dummies are weakly significant. For the sake of clarity, estimation results on the constant and the dummy variables are not shown. As is seen all factor loadings follow AR(2) processes. There are also a number of remarkable cross dynamics: first order lags in the level dynamics, $\widehat{\beta}_{i,1}$, have a positive impact on the term structure, $\widehat{\beta}_{i,3}$. Second order lags in

5.4 Semiparametric Factor Models 177

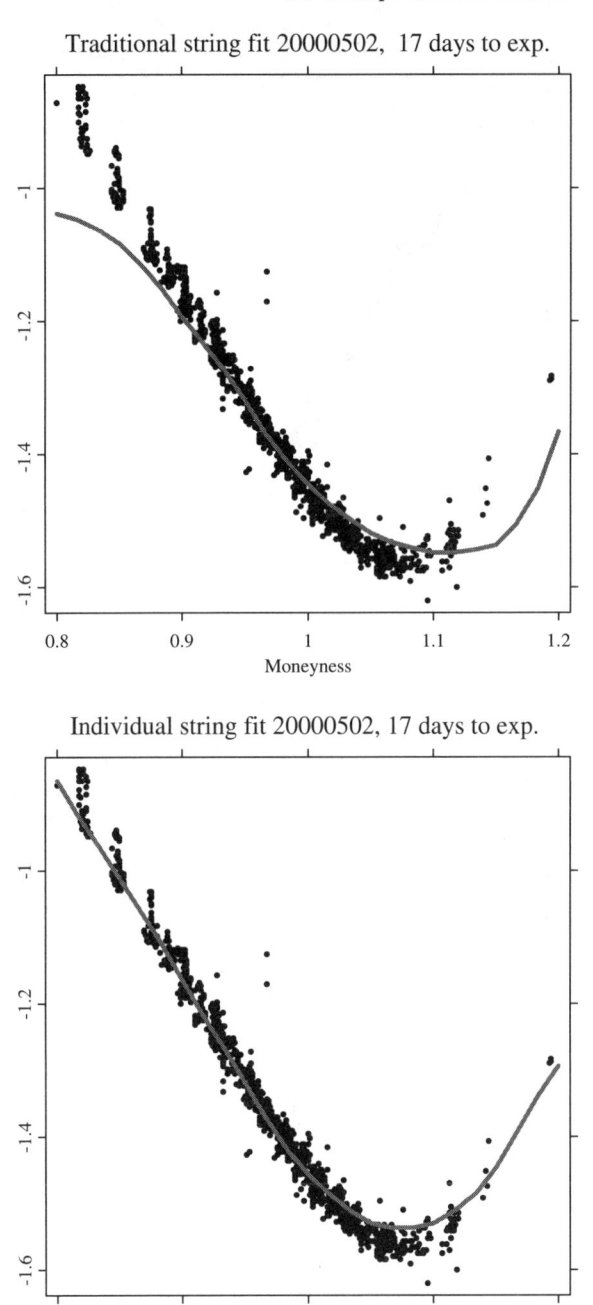

Fig. 5.21. Bias comparison of the Nadaraya-Watson estimator with $\mathbf{h} = (0.06, 0.25)^\top$ (*top panel*) and the SFM with $\mathbf{h} = (0.03, 0.04)^\top$ (*bottom panel*) for the 17 days to expiry data (*black dots*) on 20000502

178   5 Dimension-Reduced Modeling

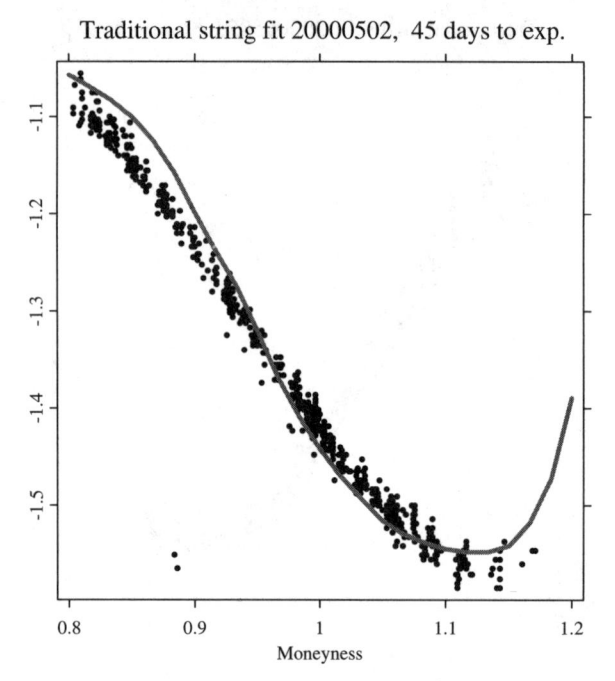

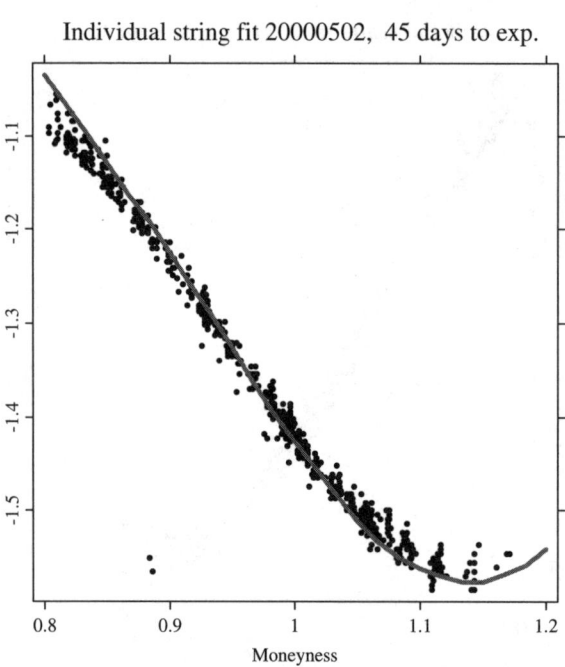

Fig. 5.22. Bias comparison of the Nadaraya-Watson estimator with $\mathbf{h} = (0.06, 0.25)^\top$ (*top panel*) and the SFM with $\mathbf{h} = (0.03, 0.04)^\top$ (*bottom panel*) for the 45 days to expiry data (*black dots*) on 20000502

5.4 Semiparametric Factor Models 179

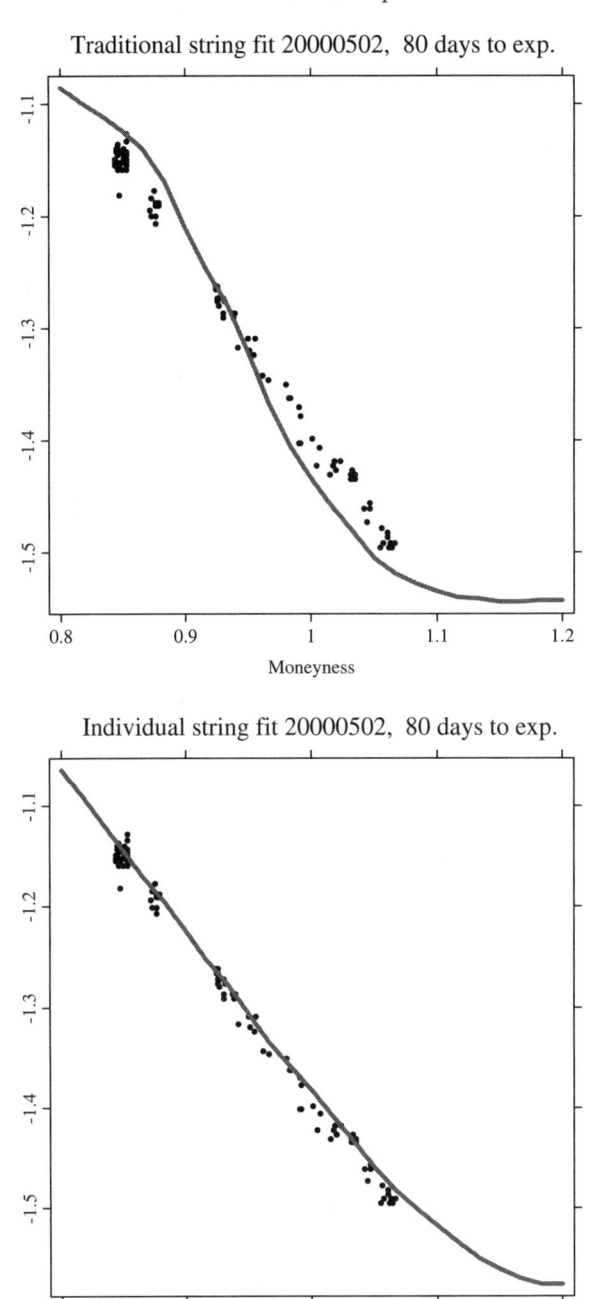

Fig. 5.23. Bias comparison of the Nadaraya-Watson estimator with $\mathbf{h} = (0.06, 0.25)^\top$ (*top panel*) and the SFM with $\mathbf{h} = (0.03, 0.04)^\top$ (*bottom panel*) for the 80 days to expiry data (*black dots*) on 20000502

180   5 Dimension-Reduced Modeling

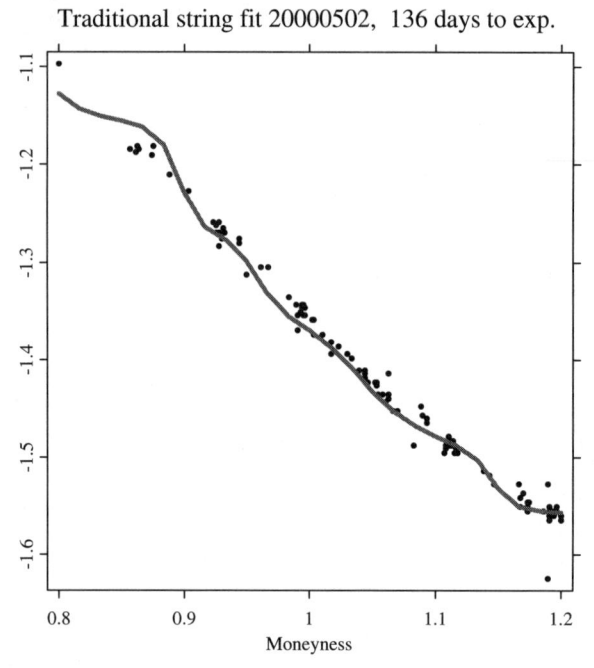

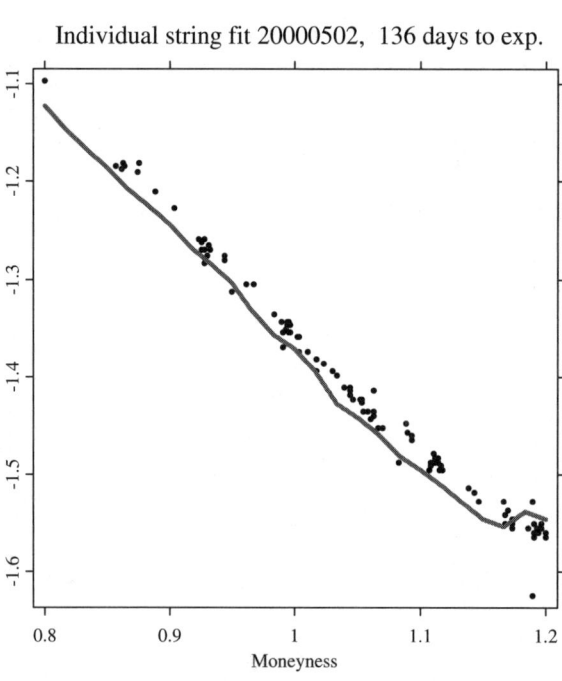

Fig. 5.24. Bias comparison of the Nadaraya-Watson estimator with $\mathbf{h} = (0.06, 0.25)^\top$ (*top panel*) and the SFM with $\mathbf{h} = (0.03, 0.04)^\top$ (*bottom panel*) for the 136 days to expiry data (*black dots*) on 20000502

5.4 Semiparametric Factor Models 181

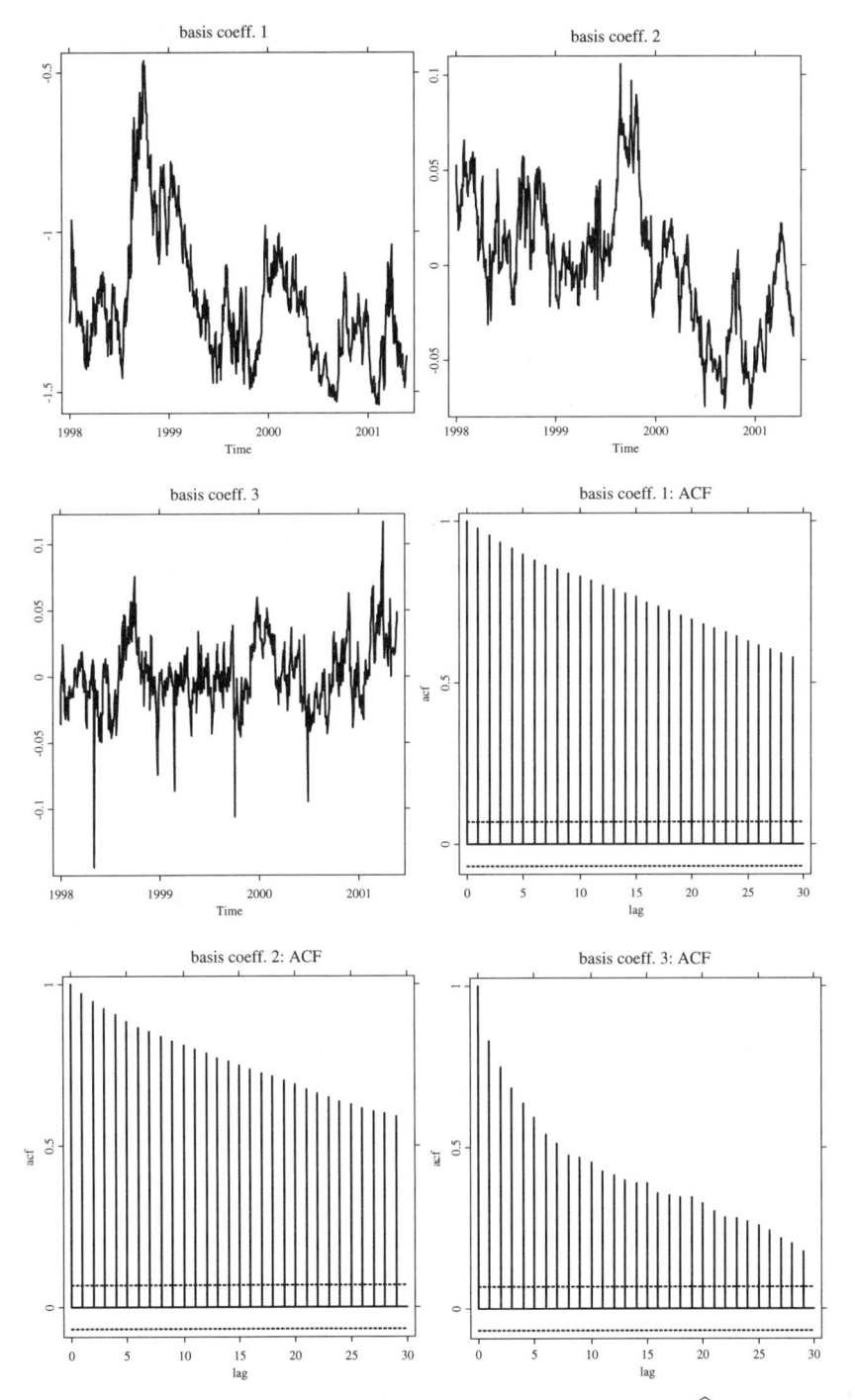

Fig. 5.25. *Upper panel and left central panels*: time series of weights $\widehat{\beta}$. *Right central panel and lower panels*: autocorrelation functions

Table 5.10. Summary statistics of SFM factor loadings $\widehat{\beta}$

	Min.	Max.	Mean	Median	Stdd.	Skewn.	Kurt.
$\widehat{\beta}_1$	−1.541	−0.462	−1.221	−1.260	0.206	1.101	4.082
$\widehat{\beta}_2$	−0.075	0.106	0.001	0.002	0.034	0.046	2.717
$\widehat{\beta}_3$	−0.144	0.116	0.002	−0.001	0.025	0.108	5.175

Table 5.11. Contemporaneous correlation matrix of $\widehat{\beta}$

	$\widehat{\beta}_{i,1}$	$\widehat{\beta}_{i,2}$	$\widehat{\beta}_{i,3}$
$\widehat{\beta}_{i,1}$	1	0.241	0.368
$\widehat{\beta}_{i,2}$		1	−0.003
$\widehat{\beta}_{i,3}$			1

Table 5.12. ADF tests on $\widehat{\beta}_{i,1}$ to $\widehat{\beta}_{i,3}$ for the full IVS model, intercept included in each case. Third column gives the number of lags included in the ADF regression. For the choice of lag length, we started with four lags, and subsequently deleted lag terms, until the last lag term became significant at least at a 5% level. MacKinnon critical values for rejecting the hypothesis of a unit root are −2.87 at 5% significance level, and −3.44 at 1% significance level

Coefficient	Test Statistic	# of lags
$\widehat{\beta}_{i,1}$	−2.68	3
$\widehat{\beta}_{i,2}$	−3.20	1
$\widehat{\beta}_{i,3}$	−6.11	2

the term structure dynamics themselves influence positively the moneyness slope effect, $\widehat{\beta}_{i,2}$, and negatively the shift variable $\widehat{\beta}_{i,1}$: thus shocks in the term structure may decrease the level of the smile and aggravate the skew. Similar interpretations can be revealed from other significant coefficients in Table 5.13.

In earlier specifications of the model, we also included contemporaneous and lagged DAX returns into the regression equation. However, the competitor in our horse-race in the following section is a simple one-step predictor without any exogenous information. Therefore, we choose a simple VAR framework without exogenous variables due to fairness.

5.4.5 Assessing Prediction Performance

We now study the prediction performance of our model compared with a benchmark model. Model comparisons that have been conducted, for instance by Bakshi et al. (1997), Dumas et al. (1998), Bates (2000), and Jackwerth and Rubinstein (2001), often show that so called 'naïve trader models' perform best or only little worse than more sophisticated models. These models used by professionals simply assert that today's IV is tomorrow's IV. There

5.4 Semiparametric Factor Models

Table 5.13. Estimation results of an VAR(2) of the factor loadings $\widehat{\beta}_i$. t-statistics given in brackets, $\bar{R}^2$ denotes the adjusted coefficient of determination. The estimation includes an intercept and two dummy variables (both not shown), which assume the value one right at those days and one day after, when the corresponding IV observations of the minimum time to maturity string (10 days to expiry) were to be excluded from the estimation of the SFM

Dependent variable	$\widehat{\beta}_{i,1}$	$\widehat{\beta}_{i,2}$	$\widehat{\beta}_{i,3}$
$\widehat{\beta}_{i-1,1}$	0.978	−0.009	0.047
	[24.40]	[−1.21]	[3.70]
$\widehat{\beta}_{i-2,1}$	0.004	0.012	−0.047
	[0.08]	[1.63]	[−3.68]
$\widehat{\beta}_{i-1,2}$	0.182	0.861	0.134
	[0.92]	[23.88]	[2.13]
$\widehat{\beta}_{i-2,2}$	−0.129	0.109	−0.126
	[−0.65]	[3.03]	[−2.01]
$\widehat{\beta}_{i-1,3}$	0.115	−0.019	0.614
	[0.97]	[−0.89]	[16.16]
$\widehat{\beta}_{i-2,3}$	−0.231	0.030	0.248
	[−1.96]	[1.40]	[6.60]
$\bar{R}^2$	0.957	0.948	0.705
F-statistic	2405.273	1945.451	258.165

are two versions: the sticky strike assumption pretends that IV is constant at fixed strikes. The sticky delta or sticky moneyness version asserts the same for IVs observed at a fixed moneyness or option delta, Derman (1999). We use the sticky moneyness model as our benchmark. There are two reasons for this choice: first, from a methodological point of view, as has been shown by Balland (2002) and Daglish et al. (2003), the sticky strike rule as an assumption on the stochastic process governing IVs, is not consistent with the existence of a smile. The sticky moneyness rule, however, can be. Second, since we estimate our model in terms of moneyness, the sticky moneyness rule is most natural.

The methodology in comparing the prediction performance is as follows: as presented earlier, the resulting times series of latent factors $\widehat{\beta}_{i,l}$ is replaced by a time series model with fitted values $\widetilde{\beta}_{i,l}(\widehat{\boldsymbol{\theta}})$ based on $\widehat{\beta}_{i',l}$ with $i' \leq i-1$, $1 \leq l \leq L$, where $\widehat{\boldsymbol{\theta}}$ is a vector of estimated coefficients seen in Table 5.13. Similarly as before, we employ an AIC based on the *fitted* values as an asymptotically unbiased estimate of the mean square prediction error.

For the model comparison, we use the criterion Ξ_{AIC_1} with $w(u) \stackrel{\text{def}}{=} 1$. Additionally we penalize the dimension of the fitted time series model $\widetilde{\beta}(\boldsymbol{\theta})$:

$$\widetilde{\Xi}_{AIC} \stackrel{\text{def}}{=} N^{-1} \sum_{i}^{I} \sum_{j}^{J_i} \left\{ y_{i,j} - \sum_{l=0}^{L} \widetilde{\beta}_{i,l}(\hat{\boldsymbol{\theta}}) \widehat{m}_l(\mathbf{x}_{i,j}) \right\}^2$$
$$\times \exp\left\{ 2\frac{L}{N} K_{\mathbf{h}}(0) \mu_\lambda + \frac{2\dim(\boldsymbol{\theta})}{N} \right\}. \quad (5.88)$$

In our case $\dim(\boldsymbol{\theta}) = 27$, since we have for three equations six VAR-coefficients plus the constant and two dummy variables.

Criterion (5.88) is compared with the squared one-day prediction error of the sticky moneyness (StM) model:

$$\Xi_{StM} \stackrel{\text{def}}{=} N^{-1} \sum_{i}^{I} \sum_{j}^{J_i} (y_{i,j} - y_{i-1,j'})^2. \quad (5.89)$$

In practice, since one hardly observes $y_{i,j}$ at the same moneyness as in $i-1$, $y_{i-1,j'}$ is obtained via a localized interpolation of the previous day's smile. Time to maturity effects are neglected, and observations, the previous values of which are lost due to expiry, are deleted from the sample.

Running the model comparison shows:

$$\Xi_{StM} = 0.00476,$$
$$\widetilde{\Xi}_{AIC} = 0.00439.$$

Thus, the model comparison reveals that the SFM is approximately 10% better than the naïve trader model. This is a substantial improvement given the high variance in IV and financial data in general. An alternative approach would investigate the hedging performance of our model compared with other models, e.g. in following Engle and Rosenberg (2000). This is left for further research.

5.5 Summary

This chapter is divided into two main parts. In the first part, we presented CPC models as a natural means of modeling the IVS. The CPC approach comprises an entire hierarchy of models. This allows for a detailed analysis of the 'degree of commonness' within different maturity groups of the IVS. We derived tests to assess stability of the factor loadings across different samples and found that only the first component may be considered as being sufficiently stable. The other components fluctuate from sample to sample year. Finally, we modelled the resulting time series by means of ARCH and GARCH processes.

In the second part, we digressed on FPCA for IVS modeling. Then, we presented a semiparametric factor model as a new modeling approach to the IVS. The key advantage is that it takes care of the discrete string structure of

IV data. The technique can be seen as a combination from FPCA and backfitting in additive models. Unlike other studies, this ansatz is tailored to the degenerated design of IV data by fitting basis functions in the local neighborhood of the design points only. This can reduce bias effects in the estimation of the IVS. Due to its flexible semiparametric structure, the SFM may also be advantageous compared to the CPC approach given the structural shifts in the underlying data. After estimating the factor functions, we fitted vector autoregressive processes of order two to the factor series. The presentation of the SFM concluded with a horse race between the SFM and the 'naïve trader model'. We found the SFM to be approximately 10% superior to the more simple model.

Our analysis has shown that CPC and SFM models are powerful dimension reduction techniques in the context of IVS modeling. Typically, the IVS allows for a decomposition into three factor that drive the surface. These factors can be interpreted as a shift factor, which accounts for around 80% of the variation, a slope and a twist or term structure factor. This result can have numerous applications: an obvious one is risk management, for instance in scenario analysis and stress tests of portfolios. In order to make the SFM more tractable, it may be good to replace the nonparametric functions by suitable parametric approximations. Then, Monte Carlo simulations of the models along the lines of Jamshidian and Zhu (1997) are straightforward.

6

Conclusion and Outlook

The implied volatility (IV) smile and implied volatility surface (IVS) are empirical phenomena that have spurred research since the discovery of the Black-Scholes (BS) formula in the nineteen-seventies. Two main strands of literature have dominated the research agenda since then. The first tries to exploit IV as a predictor for asset price fluctuations. The second seeks to provide alternative option pricing models that explain the existence of the volatility smile. Recently, a third line of research has emerged: shaped by the establishment of organized futures markets that allow trading of standardized derivatives at low costs with high liquidity, this new research aims at exploiting the information content of option prices or the IVS for the pricing of more complicated derivatives or positions. This approach has been termed *smile consistent modeling*.

The IVS is an input factor in almost any smile consistent model, either directly or in some intermediate step such as the reconstruction of the local volatility surface: it may come along as a simple estimate of the current surface or as a fully specified dynamic model describing the propagation of the IVS through time. Its accuracy and precision are the decisive competitive advantages for any smile consistent pricing model. This is particularly obvious for the complex derivatives and structured products that emerged on the markets: several underlying assets of all different kinds such as stocks, bonds and commodity linked products are comprised into a single structured derivative with complicated path-dependent payoffs, Overhaus (2002) and Quessette (2002). These products are likely to exhibit high sensitivity to volatility and are very susceptible to any misspecification of the volatility process.

Besides introducing into the financial theory of smile consistent approaches, the aim of this book is to take a specific semiparametric perspective towards two main aspects of model building of the IVS: smoothing and dimension-reduced modeling. We believe that such an approach is well placed given the challenges we face in this context: the unknown, complicated functional form of the IVS and its intricate discrete design. Non- and semiparametric techniques do not require any a priori knowledge of the functional form which

6 Conclusion and Outlook

is fitted to the data. Rather, it is the IV observations that 'decide'. Since from theory only loose restrictions on the IVS can be derived, for instance in terms of wide no-arbitrage bounds on the slopes, this approach appears to be particularly attractive.

Smile consistent models are a fruitful field of research, and we can resort on a wide spectrum of different approaches and specifications today. However, the current literature lacks empirical assessments and especially investigations of their hedging performance. These studies should include exotic options and be performed in comparison with competing model classes, such as stochastic volatility and jump-diffusion models. This will also shed new light on the delta debate. Stochastic variants of local volatility models may serve as an elegant way to circumvent the delta problem, Derman and Kani (1998), Alexander and Nogueira (2004), but it remains to be shown how they can be employed effectively for the pricing of exotic derivatives.

A topic for further research is the stability of the dimension reduction. Instead of estimating on predefined intervals, an alternative is to embed it into a framework of adaptive window choice as developed by Spokoiny (1998). Within this setting, one would aim at identifying time-homogeneous intervals on which the dimension reduction is performed. Examples of this approach in (realized) volatility modeling are Härdle et al. (2003), Mercurio (2004), and Mercurio and Spokoiny (2004).

Besides from modeling the IVS, common principle component (CPC) models are a natural choice whenever the data fall into a number of groups. This is encountered a lot of times in economics and finance: for instance, the same variables may be measured in different countries and markets. Thus, CPC models have found application in the analysis of the term structure of interest rates across different countries, Alexander and Lvov (2003) and Pérignon and Villa (2002, 2004). Other possible applications are obvious. Similar reflections apply to the semiparametric factor model (SFM). Its main properties – estimation in the local neighborhood of the design points and suitable dimension reduction – make it an ideal candidate for functional modeling. Potential fields of application are the term structure of interest rates, or swap and forward rates.

We believe that semiparametric modeling in finance is an inspiring field of research, and – in recalling the words of Corrozet (1543) – it appears to be particularly fruitful in a financial world that is 'un monde instable porté sur la mer tant esmeue et rogue'.

A

Description and Preparation of the IV Data

A.1 Preliminaries

The data set employed for this research contains tick statistics on the DAX futures contract and DAX index options and is provided by the EUREX (Frankfurt am Main) for the period from 19950101 to 20010531. Both futures contract data and option data are contract based data, i.e. each single contract is registered together with its price, contract size, and time of settlement up to a hundredth second. Interest rate data in daily frequency, i.e. one, three, six and twelve months FIBOR rates for the years 1995–1999 and EURIBOR rates for the period 2000–2001, are obtained from *Thomson Financial Datastream*. Interest rate data are linearly interpolated to approximate the riskless interest rate for the option's time to maturity. In order to avoid a German tax bias, option raw data has undergone a preparation scheme which is due to Hafner and Wallmeier (2001) and described in the following. The entire data set is stored in the financial database MD*base, maintained at the Center for Applied Statistics and Economics (CASE) at the Humboldt-Universität zu Berlin.

It is important to remark that a number of fundamental amendments in income taxation were introduced in Germany in 2000 (Steuersenkungsgesetz, BGBl. Teil I, Nr. 46 dating from 20001026). After a transition period starting in 2001, the changes came fully into effect beginning from 2002. The former legislation granted a tax voucher to domestic shareholders in compensation for the corporate tax paid by the company (Anrechnungsverfahren). However, this did not apply to foreign investors. Since 2002, the taxes paid on corporate income can no longer be deducted by domestic shareholders. Instead, 50% of the distributed dividends are taxed at the personal income tax (Halbeinkünfteverfahren), while the other 50% of the capital income are not liable to any further taxation. Therefore, the correction may no longer be mandatory for the DAX index option data beginning from 2002. Regrettably, we are not aware of any study investigating this issue. For details on German taxation law, we refer for instance to Tipke et al. (2002) or Rose (2004).

A.2 Data Correction Scheme

In a first step of the correction scheme, the DAX index values are recovered. To this end, we group to each option price observation H_t the futures price F_t of the nearest available futures contract, which was traded within a one minute interval around the observed option. The futures price observation was taken from the most heavily traded futures contract on the particular day, which is the three months contract. The no-arbitrage price of the underlying index in a frictionless market without dividends is given by

$$S_t = e^{-r_{T_F,t}(T_F-t)} F_t \,, \qquad (A.1)$$

where S_t and F_t denote the index and the futures price respectively, T_F the maturity date of the futures contract, and $r_{T,t}$ the interest rate with maturity $T-t$.

The DAX index is a capital weighted performance index, Deutsche Börse (2002), i.e. dividends less corporate tax are reinvested into the index. Therefore, at a first glance, dividend payments should have no or almost little impact on the index options. However, when only the interest rate discounted futures price is used to recover IVs by inverting the BS formula, IVs of calls and puts can differ significantly. This discrepancy is especially large during spring, when most of the 30 companies listed in the DAX distribute dividends. The point is best visible in Fig. A.1 from 20000404: IVs of calls (crosses) and

Implied Volatility Surface Ticks

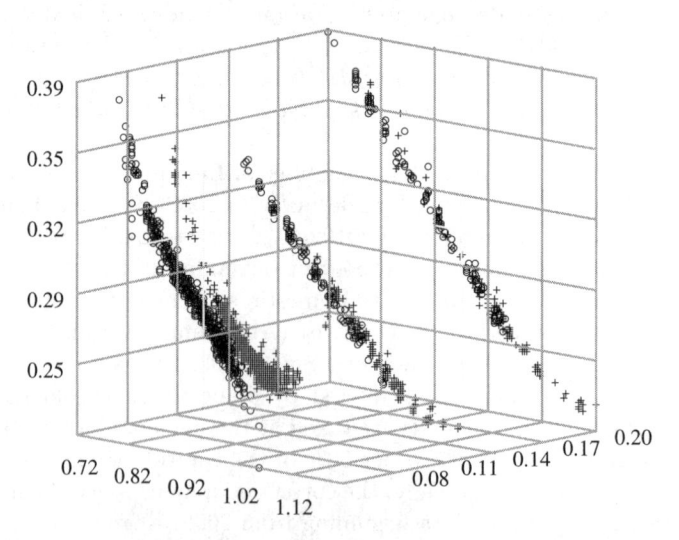

Fig. A.1. IVS ticks on 20000404, derived from futures prices that are interest rate discounted only. Put IV are *circles*, call IV *crosses*

Implied Volatility Surface Ticks

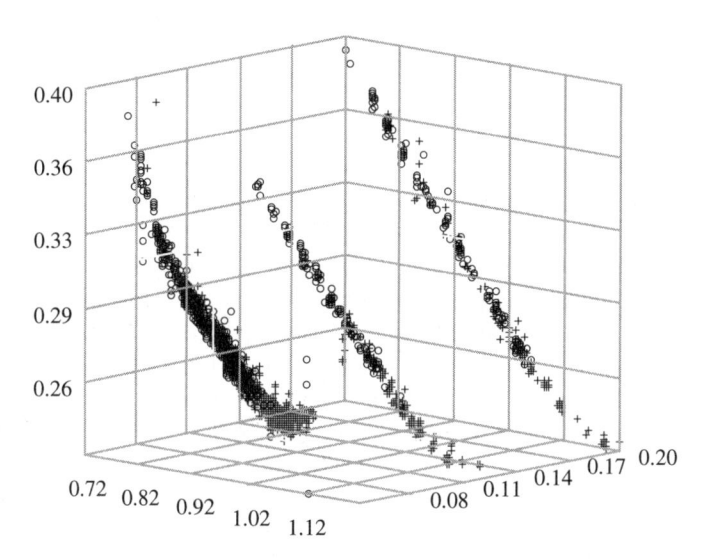

Fig. A.2. IVS ticks on 20000404, derived from futures prices that are interest rate discounted *and* corrected with the implied difference dividend. Put IV are *circles*, call IV *crosses*

puts (circles) fall apart, thus violating the put-call-parity (2.26) and general market efficiency considerations.

Hafner and Wallmeier (2001) argue that the marginal investor's individual tax scheme is different from the one actually assumed to compute the DAX index. As has been explained in Sect. A.1, this can be the case between foreign and domestic shareholders, or between domestic shareholders of different individual taxation. Consequently, the net dividend for this investor can be higher or lower than the one used for the index computation. The discrepancy, which the authors call *difference dividend*, has the same impact as a dividend payment for an unprotected option, i.e. it drives a wedge into the option prices and hence into IVs. Denote by $\Delta D_{t,T}$ the time T value of this difference dividend incurred between t and T. Consider the dividend adjusted futures price, which is approximated here by the forward price:

$$F_t = e^{r_F(T_F-t)} S_t - \Delta D_{t,T_F} , \tag{A.2}$$

and the dividend adjusted put-call parity:

$$C_t - P_t = S_t - \Delta D_{t,T_H} e^{-r_H(T_H-t)} - e^{-r_H(T_H-t)} K , \tag{A.3}$$

with T_H denoting the call's C_t and the put's P_t maturity date. Inserting (A.2) into (A.3) yields

$$C_t - P_t = F_t e^{-r_F(T_F-t)} + \Delta D_{t,T_H,T_F} - e^{-r_H(T_H-t)} K, \qquad (A.4)$$

where $\Delta D_{t,T_H,T_F} \stackrel{\text{def}}{=} \Delta D_{t,T_F} e^{-r_F(T_F-t)} - \Delta D_{t,T_H} e^{-r_H(T_H-t)}$ is the desired difference dividend.

The 'adjusted' index level

$$\tilde{S}_t = F_t e^{-r_F(T_F-t)} + \Delta D_{t,T_H,T_F} \qquad (A.5)$$

is that index level, which ties put and call IVs exactly to the same levels when used in the inversion of the BS formula.

For an estimate of $\Delta \hat{D}_{t,T_H,T_F}$, pairs of puts and calls of the strikes and same maturity are identified provided they were traded within a five minutes interval. For each pair the $\Delta D_{t,T_H,T_F}$ is derived from (A.4). To ensure robustness $\Delta \hat{D}_{t,T_H,T_F}$ is estimated by the median of all $\Delta D_{t,T_H,T_F}$ of the pairs for a given maturity at day t. IVs are recovered by inverting the BS formula using the corrected index value $\tilde{S}_t = F_t e^{-r_F(T_F-t)} + \Delta \hat{D}_{t,T_H,T_F}$. Note that $\Delta D_{t,T_H,T_F} = 0$, when $T_H = T_F$. Indeed, when calculated also in this case, $\Delta \hat{D}_{t,T_H,T_F}$ proved to be very small (compared with the index value), which supports the validity of this approach. The described procedure is applied on a daily basis throughout the entire data set from 19950101 to 20010531. All computations have been made with XploRe, Härdle et al. (2000b).

In Fig. A.2, also from 20000404, we present the data after correcting the discounted futures price with an implied difference dividend $\Delta \hat{D}_t = (10.3, 5.0, 1.9)^\top$, where the first entry refers to 16 days, the second to 45 days and the third to 73 days to maturity. IVs of puts and calls converge two one single string, while the concavity of the put volatility smile is remedied, too. Note that the overall level of the IV string is *not* altered through that procedure.

The data are transaction based and may contain potential misprints and outliers. This is seen in Figs. A.1 and A.2. To accommodate for this, a mild filter is applied: observations with IV less than 4% and bigger than 80% are dropped. Furthermore, we disregard all observations having a maturity $\tau \leq 10$ days. Obviously, this filter does not detect outliers within these bounds. At this point *robust statistical methods* may be an adequate choice. However, given the sheer vastness of the data set, we believe this filter still to be adequate.

After this filtering, the entire number of observations are more than 5.7 million contracts. Trading volume increased considerably during this sample period. For the last years 1998, and the following years it is around 5 200 observations per day.

Table A.1 gives a short summary of our IVS data. Most heavy trading occurs in the short term contracts, as is seen from the difference between median and mean of the term structure distribution of the observations as

Table A.1. Summary statistics on the data base from 19950101 to 20010531, entirely and on an annual basis. 2001 is from 20010101 to 20010531, only

		Min.	Max.	Mean	Median	Stdd.	Skewn.	Kurt.
All	T. to mat.	0.028	2.014	0.134	0.084	0.149	3.623	22.574
	Moneyn.	0.325	1.856	0.987	0.994	0.097	−0.303	5.801
	IV	0.041	0.799	0.255	0.246	0.088	1.531	7.531
1995	T. to mat.	0.028	0.769	0.132	0.086	0.121	2.265	8.441
	Moneyn.	0.771	1.207	0.996	0.997	0.040	−0.111	4.530
	IV	0.046	0.622	0.149	0.147	0.021	1.218	12.165
1996	T. to mat.	0.028	2.011	0.152	0.097	0.167	3.915	28.561
	Moneyn.	0.687	1.221	0.987	0.993	0.044	−0.723	5.887
	IV	0.046	0.789	0.134	0.130	0.028	2.893	33.466
1997	T. to mat.	0.028	1.964	0.147	0.086	0.172	3.503	21.267
	Moneyn.	0.446	1.441	0.979	0.988	0.077	−0.546	5.442
	IV	0.043	0.800	0.246	0.233	0.073	1.149	5.027
1998	T. to mat.	0.028	2.014	0.134	0.081	0.148	3.548	22.957
	Moneyn.	0.386	1.856	0.984	0.992	0.108	−0.030	5.344
	IV	0.041	0.799	0.335	0.306	0.114	0.970	3.471
1999	T. to mat.	0.028	1.994	0.126	0.083	0.139	4.331	32.578
	Moneyn.	0.371	1.516	0.979	0.992	0.099	−0.595	5.563
	IV	0.047	0.798	0.273	0.259	0.076	0.942	4.075
2000	T. to mat.	0.028	1.994	0.130	0.083	0.151	3.858	23.393
	Moneyn.	0.325	1.611	0.985	0.992	0.092	−0.337	6.197
	IV	0.041	0.798	0.254	0.242	0.060	1.463	7.313
2001	T. to mat.	0.028	0.978	0.142	0.083	0.159	2.699	10.443
	Moneyn.	0.583	1.811	1.001	1.001	0.085	0.519	6.762
	IV	0.043	0.789	0.230	0.221	0.049	1.558	7.733

well as from its skewness. Median time to maturity is 30 days (0.083 years). Across moneyness the distribution is slightly negatively skewed. Mean IV over the sample period is 27.9%.

B

Some Results from Stochastic Calculus

This chapter contains a number of basic definitions and results from stochastic calculus. They are collected in order to make our treatment more self-contained. Thus, the selection of the issues is driven by their complementary function to our work, rather than by their importance in stochastic calculus. For any deeper treatment or proofs, we refer to standard textbooks such as Øksendal (1998), Karatzas and Shreve (1991), or Steele (2000).

In this chapter, we consider stochastic processes defined on a complete probability space $(\Omega, \mathcal{F}, \mathrm{P})$. The probability space is equipped with a filtration, i.e. a nondecreasing family $(\mathcal{F}_t)_{t \geq 0}$ of subsigma fields $\mathcal{F}_s \subseteq \mathcal{F}_t \subseteq \mathcal{F}$, for $0 \leq s < t$. The filtration is assumed to satisfy the 'usual' conditions, namely that it is right-continuous, and that $\mathcal{F}_0$ contains all null sets. A stochastic process is a collection of random variables $(X_t)_{t \geq 0}$ on $(\Omega, \mathcal{F})$, which take values in $\mathbb{R}^d$. The index t is interpreted as 'time'. We say that a stochastic process X is adapted to $(\mathcal{F}_t)_{t \geq 0}$, if all X_t are $(\mathcal{F}_t)_{t \geq 0}$-measurable. For a fixed $\omega \in \Omega$, the mapping $t \to X_t(\omega)$ for $t \geq 0$ is called the *sample path* of X associated with ω.

Martingale

Let $(X_t)_{0 \leq t < \infty}$ be an $(\mathcal{F}_t)_{t \geq 0}$-adapted stochastic process on $(\Omega, \mathcal{F}, \mathrm{P})$ satisfying $\mathsf{E}|X_t| < \infty$ for all $0 \leq t < \infty$. The process X is called an $(\mathcal{F}_t)_{t \geq 0}$-martingale, if for every $0 \leq s < t < \infty$, we have

$$\mathsf{E}(X_t | \mathcal{F}_s) = X_s \, . \tag{B.1}$$

The Quadratic Variation and Covariation Process

Let $(X_t)_{0 \leq t \leq T}$, for $T < \infty$, be an $(\mathcal{F}_t)_{t \geq 0}$-adapted stochastic process on $(\Omega, \mathcal{F}, \mathrm{P})$. Further, let $\mathcal{D}_n$ be the Dyadic decomposition of order n on the interval $[0, T]$, i.e.

196   B Some Results from Stochastic Calculus

$$\mathcal{D}_n = \{i2^{-n} | i = 0, 1, 2, 3, \ldots\} \cap [0, T] \,. \tag{B.2}$$

The quadratic variation process of X is defined by (provided it exists):

$$\langle X \rangle_t \stackrel{\text{def}}{=} \lim_{n \uparrow \infty} \sum_{0 < t_i \le t} (X_{t_i} - X_{t_{i-1}})^2 \,, \quad \text{for } 0 \le t \le T, \tag{B.3}$$

where the limit is understood in probability.

Let $(Y_t)_{0 \le t < T}$ be a second stochastic process on $(\Omega, \mathcal{F}, \mathrm{P})$. The covariation process of X and Y is defined by (if it exists):

$$\langle X, Y \rangle_t \stackrel{\text{def}}{=} \lim_{n \uparrow \infty} \sum_{0 < t_i \le t} (X_{t_i} - X_{t_{i-1}})(Y_{t_i} - Y_{t_{i-1}}), \quad \text{for } 0 \le t \le T, \tag{B.4}$$

where the limit is understood in probability.

Brownian Motion

A real-valued stochastic process $(W_t)_{0 \le t \le T < \infty}$ adapted to $(\mathcal{F}_t)_{0 \le t < T}$ is called a *standard Brownian motion* with respect to $(\mathcal{F}_t)_{0 \le t < T}$ on the interval $[0, T]$ if it satisfies the following properties:

(i) $W_0 = 0$
(ii) For any $0 \le s < t \le T$ the increment

$$W_t - W_s \tag{B.5}$$

is independent of $\mathcal{F}_s$ and has the Gaussian distribution $N(0, t - s)$.
(iii) $(W_t)_{0 \le t \le T}$ has continuous sample paths.

For $0 \le s < t \le T$ the covariance is calculated as $\mathsf{Cov}(W_s, W_t) = \mathsf{E}\{(W_t - W_s + W_s)W_s\} = \mathsf{E}(W_s^2)$, so in general

$$\mathsf{Cov}(W_s, W_t) = \min(s, t) \quad \text{for } 0 \le s, t \le T. \tag{B.6}$$

For almost every $\omega \in \Omega$ the Brownian sample path associated with ω is nowhere differentiable. However, its quadratic variation process exists and is P-almost surely:

$$\langle W \rangle_t = t, \quad \text{for } 0 \le t \le T. \tag{B.7}$$

Itô Formula

Suppose that the real-valued process X taking values in $\mathbb{R}$ has the (stochastic) integral representation

$$X_t = x_0 + \int_0^t a_s \, ds + \int_0^t b_s \, dW_s \tag{B.8}$$

on $0 \leq t \leq T$, where $(a_t)_{0 \leq t \leq T}$ and $(b_t)_{0 \leq t \leq T}$ are real-valued $(\mathcal{F}_t)_{0 \leq t \leq T}$-adapted processes satisfying

$$P\left(\int_0^T |a_s|\, ds < \infty\right) = 1 \quad \text{and} \quad P\left(\int_0^T b_s^2\, ds < \infty\right) = 1.$$

Then X is called an *Itô process*. Its quadratic variation process exists and is given by:

$$\langle X \rangle_t = \int_0^t b_s^2\, ds \tag{B.9}$$

for $0 \leq t \leq T$.

Let $f \in \mathcal{C}^{2,1}(\mathbb{R} \times \mathbb{R}^+)$. Then Itô's formula states

$$f(X_t, t) = f(X_0, 0) + \int_0^t \frac{\partial f(X_s, s)}{\partial t}\, ds + \int_0^t \frac{\partial f(X_s, s)}{\partial x}\, dX_s$$
$$+ \frac{1}{2} \int_0^t \frac{\partial^2 f(X_s, s)}{\partial x^2}\, d\langle X \rangle_s, \tag{B.10}$$

for $0 \leq t \leq T$.

For the vector-valued process $X = (X^{(1)}, \ldots, X^{(d)})^\top$ and $f \in \mathcal{C}^{2,1}(\mathbb{R}^d \times \mathbb{R}^+)$, Itô's formula generalizes to

$$f(X_t, t) = f(X_0, 0) + \int_0^t \frac{\partial f(X_s, s)}{\partial t}\, ds + \sum_{i=1}^d \int_0^t \frac{\partial f(X_s, s)}{\partial x_i}\, dX_s^{(i)}$$
$$+ \frac{1}{2} \sum_{i=1}^d \sum_{j=1}^d \int_0^t \frac{\partial^2 f(X_s, s)}{\partial x_i \partial x_j}\, d\langle X^{(i)}, X^{(j)} \rangle_s. \tag{B.11}$$

Tanaka-Meyer Formula

The Itô formula can be generalized to convex functions f, Tanaka (1963), Meyer (1976), in which case it is known as *Tanaka-Meyer formula*, Karatzas and Shreve (1991, Theorem 3.6.22 and p. 220).

For some $c \in \mathbb{R}$ consider the convex function $f : \mathbb{R} \to \mathbb{R}, x \to (x-c)^+$, which is the relevant special case in this book. The left side derivative of f is given by

$$D^- f(x) = \mathbf{1}(x > c), \tag{B.12}$$

where $\mathbf{1}(\mathcal{A})$ denotes the indicator function of the set $\mathcal{A}$.

Define the second derivative in a distributional sense by

$$\frac{\partial^2 f(x)}{\partial x^2} = \delta_c(x), \tag{B.13}$$

where δ_c is the Dirac delta function centered at c.

198   B Some Results from Stochastic Calculus

Let X satisfy representation (B.8). The Tanaka-Meyer formula states:

$$(X_t - c)^+ = (x_0 - c)^+ + \int_0^t \mathbf{1}(X_s > c)\, dX_s + \frac{1}{2} L_t^c ,\qquad \text{(B.14)}$$

for $0 \leq t \leq T$.

$$L_t^c \stackrel{\text{def}}{=} \lim_{n \uparrow \infty} \int_0^t n\, \mathbf{1}\left\{X_s \in \left(c, c + \frac{1}{n}\right)\right\} d\langle X \rangle_s \qquad \text{(B.15)}$$

$$= \int_0^t \delta_c(X_s)\, b_s^2\, ds \qquad \text{(B.16)}$$

is called the *local time* at level c. Intuitively, it measures the 'time spent at level c'.

Uniqueness and Existence of SDE

In the following, we shall denote by $(\mathcal{F}_t)_{0 \leq t \leq T}$ the P-augmentation of the filtration

$$\mathcal{F}_t^W = \sigma(W_s,\ 0 \leq s \leq t),\quad 0 \leq t \leq T , \qquad \text{(B.17)}$$

generated by W. It can be shown that $(\mathcal{F}_t)_{0 \leq t \leq T}$ is already right-continuous and thus satisfies the 'usual' conditions.

For $x_0 \in \mathbb{R}$, consider the one-dimensional SDE:

$$dX_t = a(X_t, t)\, dt + b(X_t, t)\, dW_t , \qquad \text{(B.18)}$$

with initial condition $X_0 = x_0$, and with functions $a, b : \mathbb{R} \times [0, T] \to \mathbb{R}$. Assume that they satisfy the global Lipschitz condition:

$$|a(x, t) - a(y, t)| + |b(x, t) - b(y, t)| \leq K|x - y| , \qquad \text{(B.19)}$$

and the linear growth condition:

$$|a(x, t)| + |b(x, t)| \leq L(1 + |x|) , \qquad \text{(B.20)}$$

for any $0 \leq t \leq T$, and $x, y \in \mathbb{R}$, where K, L are a positive constants. Then there exists a *strong* solution to (B.18), i.e. there exists a continuous $(\mathcal{F}_t)_{0 \leq t \leq T}$-adapted process $(X_t)_{0 \leq t \leq T}$ satisfying (B.18) and the initial condition $X_0 = x_0$.

Moreover, if $(Y_t)_{0 \leq t \leq T}$ is another solution to (B.18), then strong uniqueness holds, i.e.

$$P(X_t = Y_t \text{ for all } t \in [0, T]) = 1 . \qquad \text{(B.21)}$$

This is the one-dimensional version of, e.g., Karatzas and Shreve (1991, Theorem 5.2.9). In the vector-valued case, the absolute value is to be replaced by a norm, but similar results hold.

Fokker-Planck Equation

Let $(X_t)_{0 \leq t \leq T}$ which takes values in $\mathbb{R}$ satisfy the SDE

$$dX_t = a(X_t, t)\, dt + b(X_t, t)\, dW_t\,, \tag{B.22}$$

with initial condition $X_0 = x_0$. Under the ellipticity condition $b^2 \geq \epsilon > 0$, X is a Markov process and its transition kernel takes the form

$$\mathrm{P}(X_T \in dy | X_t = x) = \phi(y, T | X_t, t)\, dy \tag{B.23}$$

for some jointly measurable density function $\phi(y, T | X_t, t) \geq 0$. The notation makes precise that it is a density conditional on X_t and t. Then, $\phi(y, T | X_t, t)$ can be characterized by the *Fokker-Planck* or *forward Kolmogorov* equation

$$0 = \frac{\partial \phi(y, T | X_t, t)}{\partial T} + \frac{\partial \{a(y, T) \phi(y, T | X_t, t)\}}{\partial y} - \frac{1}{2} \frac{\partial^2 \{b^2(y, T) \phi(y, T | X_t, t)\}}{\partial y^2} \tag{B.24}$$

for fixed $(X_t, t) \in \mathbb{R} \times \mathbb{R}^+$ and with the initial condition

$$\phi(y, t | X_t, t) = \delta_y(X_t)\,. \tag{B.25}$$

Girsanov's Theorem

Let $W = (W^{(1)}, \ldots, W^{(d)})^\top$ be a d-dimensional standard Brownian motion defined on Ω and $0 \leq T < \infty$. Further let $\alpha = (\alpha^{(1)}, \ldots, \alpha^{(d)})$ be an $\mathbb{R}^d$-valued $(\mathcal{F}_t)_{0 \leq t \leq T}$-adapted process which satisfies $\mathrm{P}\{\int_0^T (\alpha_s^{(i)})^2\, ds < \infty\} = 1$ for each $i = 1, \ldots, d$.

Define the process

$$M_t \stackrel{\text{def}}{=} \exp\left(\sum_{i=1}^d \int_0^t \alpha_s^{(i)}\, dW_s^{(i)} - \frac{1}{2} \int_0^t \|\alpha_s^{(i)}\|^2\, ds\right), \tag{B.26}$$

where $\|\cdot\|$ denotes the Euclidian norm. Assume that α satisfies the *Novikov* condition:

$$\mathsf{E}\left\{\exp\left(\frac{1}{2}\int_0^T \|\alpha_s\|^2\, ds\right)\right\} < \infty\,. \tag{B.27}$$

Then $(M_t)_{0 \leq t \leq T}$ is a martingale and

$$\mathsf{E} M_t = 1\,, \tag{B.28}$$

for each $0 \leq t \leq T$. Thus, we can define a new probability measure $\widetilde{\mathrm{P}}_T$ on $(\Omega, \mathcal{F}_T)$ by

$$\widetilde{\mathrm{P}}_T(A) \stackrel{\text{def}}{=} \mathsf{E}\{\mathbf{1}(A) M_T\}\,, \quad A \in \mathcal{F}_T\,, \tag{B.29}$$

i.e. $\widetilde{\mathrm{P}}$ has the Radon-Nikodým derivative:

$$\frac{d\widetilde{\mathrm{P}}_T}{d\mathrm{P}} = M_T . \tag{B.30}$$

We can also define a new process $\widetilde{W} = (\widetilde{W}^{(1)}, \ldots, \widetilde{W}^{(d)})^\top$ by

$$\widetilde{W}_t^{(i)} \stackrel{\text{def}}{=} W_t^{(i)} - \int_0^t \alpha_s^{(i)} \, ds , \tag{B.31}$$

for $i = 1, \ldots, d$ and $0 \leq t \leq T$.

In this situation Girsanov's theorem asserts that $\widetilde{W}$ is a standard Brownian motion on the new probability space $(\Omega, \mathcal{F}, \widetilde{\mathrm{P}}_T)$.

C

Proofs of the Results on the LSK IV Estimator

As mentioned in Sect. 4.5, proofs of these results in the general class of kernel M-estimators are due to Gouriéroux et al. (1994), here given as in Fengler and Wang (2003).

C.1 Proof of Consistency

For notational simplicity, we introduce:

$$Z(x,y) \stackrel{\text{def}}{=} w(x)\, K_{(1)}\left(\frac{\kappa_t - x}{h_{1,n}}\right) K_{(2)}\left(\frac{\tau - y}{h_{2,n}}\right), \quad \text{(C.1)}$$

and

$$\widehat{L}_n(\sigma) \stackrel{\text{def}}{=} \frac{1}{nh_{1,n}h_{2,n}} \sum_{i=1}^{n} \{\widetilde{c}_{t_i} - c^{BS}(\kappa_{t_i}, \tau_i, r_i, \sigma)\}^2 \, Z(\kappa_{t_i}, \tau_i). \quad \text{(C.2)}$$

and we remind that throughout this chapter $\kappa_t \stackrel{\text{def}}{=} K/S_t$. For sake of clarity, we drop in the following the explicit dependence of the option prices and its derivatives on r. Moreover, in this and the following section E_t is an abbreviation for the conditional expectation with respect to $\mathcal{F}_t$.

As a first step, let us prove

$$\widehat{L}_n(\sigma) \xrightarrow{p} L(\sigma) \stackrel{\text{def}}{=} \mathsf{E}_t\left[\{\widetilde{c}_t - c^{BS}(\kappa_t, \tau, \sigma)\}^2 w(\kappa_t)\right]. \quad \text{(C.3)}$$

It is observed that

$$\begin{aligned}
\widehat{L}_n(\sigma) &= \frac{1}{nh_{1,n}h_{2,n}} \sum_{i=1}^{n} \big\{ \{\widetilde{c}_{t_i} - c^{BS}(\kappa_{t_i}, \tau_i, \sigma)\}^2 \, Z(\kappa_{t_i}, \tau_i) \\
&\quad - \mathsf{E}_t\left[\{\widetilde{c}_{t_i} - c^{BS}(\kappa_{t_i}, \tau_i, \sigma)\}^2 \, Z(\kappa_{t_i}, \tau_i)\right]\big\} \\
&\quad + \frac{1}{h_{1,n}h_{2,n}} \mathsf{E}_t\left[\{\widetilde{c}_{t_1} - c^{BS}(\kappa_{t_1}, \tau_1, \sigma)\}^2 \, Z(\kappa_{t_1}, \tau_1)\right] \\
&\stackrel{\text{def}}{=} \alpha_n + \beta_n.
\end{aligned} \quad \text{(C.4)}$$

Standard arguments can be used to prove

$$\mathsf{E}_t \alpha_n^2 = \mathcal{O}\Big((nh_{1,n}h_{2,n})^{-1}\Big) \tag{C.5}$$

by conditions (A1) and (A2) on page 116.

By Taylor's expansion, we have

$$\begin{aligned}
\beta_n &= \frac{1}{h_{1,n}h_{2,n}} \mathsf{E}_t \int \{\tilde{c}_{t_1} - c^{BS}(x,y,\sigma)\}^2 \, Z(x,y) \, dx \, dy \\
&= \mathsf{E}_t \int \{\tilde{c}_t - c^{BS}(\kappa_t - h_{1,n}u, \tau - h_{2,n}v, \sigma)\}^2 \\
&\quad \times w(\kappa_t - h_{1,n}u) K_{(1)}(u) K_{(2)}(v) \, du \, dv \xrightarrow{p} L(\sigma).
\end{aligned} \tag{C.6}$$

Equations (C.5) and (C.6) together prove (C.3).

In a second step, we have, recalling the definition of $\sigma(\kappa_t, \tau)$:

$$\begin{aligned}
\frac{\partial L(\sigma)}{\partial \sigma}\bigg|_{\sigma=\sigma(\kappa_t,\tau)} &= -2\,\mathsf{E}_t \tilde{c}_t w(\kappa_t) \frac{\partial}{\partial \sigma} c^{BS}(\kappa_t, \tau, \sigma)\bigg|_{\sigma=\sigma(\kappa_t,\tau)} \\
&\quad + 2\,\mathsf{E}_t c^{BS}(\kappa_t, \tau, \sigma(\kappa_t,\tau)) w(\kappa_t) \frac{\partial}{\partial \sigma} c^{BS}(\kappa_t, \tau, \sigma)\bigg|_{\sigma=\sigma(\kappa_t,\tau)} \\
&= 0,
\end{aligned} \tag{C.7}$$

and

$$\begin{aligned}
\frac{\partial^2 L(\sigma)}{\partial \sigma^2}\bigg|_{\sigma=\sigma(\kappa_t,\tau)} &= -2\,\mathsf{E}_t \tilde{c}_t w(\kappa_t) \frac{\partial^2}{\partial \sigma^2} c^{BS}(\kappa_t, \tau, \sigma)\bigg|_{\sigma=\sigma(\kappa_t,\tau)} \\
&\quad + 2\,\mathsf{E}_t w(\kappa_t) \left(\frac{\partial}{\partial \sigma} c^{BS}(\kappa_t, \tau, \sigma)\bigg|_{\sigma=\sigma(\kappa_t,\tau)}\right)^2 \\
&\quad + 2\,\mathsf{E}_t w(\kappa_t) c^{BS}(\kappa_t, \tau, \sigma(\kappa,\tau)) \frac{\partial^2}{\partial \sigma^2} c^{BS}(\kappa_t, \tau, \sigma)\bigg|_{\sigma=\sigma(\kappa,\tau)} \\
&= 2\,\mathsf{E}_t w(\kappa_t) \left(\frac{\partial}{\partial \sigma} c^{BS}(\kappa_t, \tau, \sigma)\bigg|_{\sigma=\sigma(\kappa_t,\tau)}\right)^2.
\end{aligned} \tag{C.8}$$

This together with (C.3) proves that $\widehat{L}_n(\sigma)$ converges in probability to a convex function with a unique minimum at $\sigma = \sigma(\kappa_t, \tau)$. Thus, $\widehat{\sigma}_n(\kappa_t, \tau) \xrightarrow{p} \sigma(\kappa_t, \tau)$ is proved.

C.2 Proof of Asymptotic Normality

Recalling the definition of $\widehat{\sigma}(\kappa_t, \tau)$, it follows that $\widehat{\sigma}(\kappa_t, \tau)$ is the solution of the following equation:

$$U_n(\sigma) \stackrel{\text{def}}{=} \frac{1}{nh_{1,n}h_{2,n}} \sum_{i=1}^{n} A_i(\kappa_{t_i}, \tau_i, \sigma) B_i(\kappa_{t_i}, \tau_i, \sigma) \, Z(\kappa_{t_i}, \tau_i)$$
$$= 0 \,. \tag{C.9}$$

By Taylor's expansion, we get

$$0 = U_n(\widehat{\sigma}(\kappa_t, \tau)) = U_n(\sigma(\kappa_t, \tau)) + U_n'(\sigma^*)\Big(\widehat{\sigma}_t(\kappa_t, \tau) - \sigma(\kappa_t, \tau)\Big) \,, \tag{C.10}$$

where σ^* lies between σ and $\widehat{\sigma}$ and $U_n'(\sigma^*) \stackrel{\text{def}}{=} \frac{\partial}{\partial \sigma} U_n(\sigma)|_{\sigma=\sigma^*}$.
From (C.10), we have

$$\widehat{\sigma}(\kappa_t, \tau) - \sigma(\kappa_t, \tau) = -\{U_n'(\sigma^*)\}^{-1} U_n(\sigma) \,. \tag{C.11}$$

By some algebra, we obtain

$$\begin{aligned}
U_n'(\sigma) &= \frac{1}{nh_{1,n}h_{2,n}} \sum_{i=1}^{n} \bigg(\Big\{ (\frac{\partial}{\partial \sigma} A_i(\kappa_{t_i}, \tau_i, \sigma)) B_i(\kappa_{t_i}, \tau_i, \sigma) \\
&\quad + A_i(\kappa_{t_i}, \tau_i, \sigma)(\frac{\partial}{\partial \sigma} B_i(\kappa_{t_i}, \tau_i, \sigma)) \Big\} Z(\kappa_{t_i}, \tau_i) \\
&\quad - \mathsf{E}_t \Big[\Big\{ (\frac{\partial}{\partial \sigma} A_i(\kappa_{t_i}, \tau_i, \sigma)) B_i(\kappa_{t_i}, \tau_i, \sigma) \\
&\quad + A_i(\kappa_{t_i}, \tau_i, \sigma) \frac{\partial}{\partial \sigma} B_i(\kappa_{t_i}, \tau_i, \sigma) \Big\} Z(\kappa_{t_i}, \tau_i) \Big] \bigg) \\
&\quad + \frac{1}{nh_{1,n}h_{2,n}} \sum_{i=1}^{n} \mathsf{E}_t \Big[\Big\{ (\frac{\partial}{\partial \sigma} A_i(\kappa_{t_i}, \tau_i, \sigma)) B_i(\kappa_{t_i}, \tau_i, \sigma) \\
&\quad + A_i(\kappa_{t_i}, \tau_i, \sigma) \frac{\partial}{\partial \sigma} B_i(\kappa_{t_i}, \tau_i, \sigma) \Big\} Z(\kappa_{t_i}, \tau_i) \Big] \\
&\stackrel{\text{def}}{=} \Delta_{1,n} + \Delta_{2,n} \,.
\end{aligned} \tag{C.12}$$

Inspect first $\triangle_{1,n}$ in Equation (C.12): by some algebra, we get

$$E_t \triangle_{1,n}^2 \leq \frac{1}{n^2 h_{1,n}^2 h_{2,n}^2} \sum_{i=1}^n E_t \Big[\big\{ (\frac{\partial}{\partial \sigma} A_i(\kappa_{t_i}, \tau_i, \sigma)) B_i(\kappa_{t_i}, \tau_i, \sigma)$$

$$+ A_i(\kappa_{t_i}, \tau_i, \sigma) \frac{\partial}{\partial \sigma} B_i(\kappa_{t_i}, \tau_i, \sigma) \big\} Z(\kappa_{t_i}, \tau_i) \Big]^2$$

$$= \frac{f_t^2(\kappa_t, \tau) \int K_{(1)}^2(u) \, du \int K_{(2)}^2(v) dv}{n h_{1,n} h_{2,n}} E_t \Big[\big\{ (\frac{\partial}{\partial \sigma} A_1(\kappa_t, \tau, \sigma) B_1(\kappa_t, \tau, \sigma)$$

$$+ A_1(\kappa_t, \tau, \sigma) \frac{\partial}{\partial \sigma} B_1(\kappa_t, \tau, \sigma) \big)^2 \big\} w(\kappa_t) \Big]$$

$$+ o\Big(\frac{1}{n h_{1,n} h_{2,n}}\Big) \longrightarrow 0, \qquad (C.13)$$

as $n h_{1,n} h_{2,n} \to \infty$. The joint (time-$t$ conditional) probability density function of κ_t and τ is denoted by $f_t(\kappa_t, \tau)$.

To consider $\triangle_{2,n}$ in Equation (C.12), denote $D(\kappa_t, \tau, \sigma) \overset{\text{def}}{=} \frac{\partial}{\partial \sigma} B(\kappa_t, \tau, \sigma)$, for simplicity. Note that $\frac{\partial}{\partial \sigma} A(\kappa_t, \tau, \sigma) = -B(\kappa_t, \tau, \sigma)$. Thus, we have:

$$\triangle_{2,n} = \frac{1}{h_{1,n} h_{2,n}} E_t \Big\{ \int \Big(-B^2(x, y, \sigma) + A(x, y, \sigma) D(x, y, \sigma) \Big)$$

$$\times Z(x, y) f_t(x, y) \, dx \, dy \Big\}$$

$$= E_t \int \Big\{ -B^2(\kappa_t - h_{1,n} u, \tau - h_{2,n} v, \sigma)$$

$$+ A(\kappa_t - h_{1,n} u, \tau - h_{2,n} v, \sigma) D(\kappa_t - h_{1,n} u, \tau - h_{2,n} v, \sigma) \Big\}$$

$$\times w(\kappa_t) f_t(\kappa_t - h_{1,n} u, \tau - h_{2,n} v) K_{(1)}(u) K_{(2)}(v) \, du \, dv$$

$$\longrightarrow \Big[-E_t \Big\{ B^2(\kappa_t, \tau, \sigma) w(\kappa_t) \Big\}$$

$$+ E_t \Big\{ A(\kappa_t, \tau, \sigma) D(\kappa_t, \tau, \sigma) w(\kappa_t) \Big\} \Big] f_t(\kappa_t, \tau) . \qquad (C.14)$$

Equations (C.12), (C.13), (C.14) and the fact $U_n'(\sigma^*) - U_n'(\sigma) \to 0$ together prove:

$$U_n'(\sigma^*) \xrightarrow{p} \Big[E_t \Big\{ -B^2(\kappa_t, \tau, \sigma) w(\kappa_t) \Big\}$$

$$+ E_t \Big\{ A(\kappa_t, \tau, \sigma) D(\kappa_t, \tau, \sigma) w(\kappa_t) \Big\} \Big] f_t(\kappa_t, \tau) . \qquad (C.15)$$

Now, let

$$u_{ni} \overset{\text{def}}{=} \frac{1}{h_{1,n} h_{2,n}} A(\kappa_{t_i}, \tau_i, \sigma) B(\kappa_{t_i}, \tau_i, \sigma) \, Z(\kappa_{t_i}, \tau_i) . \qquad (C.16)$$

C.2 Proof of Asymptotic Normality 205

For some $\delta > 0$, we have:

$$\begin{aligned}
E_t|u_{ni}|^{2+\delta} &= \frac{1}{h_{1,n}^{2+\delta} h_{2,n}^{2+\delta}} E_t A^{2+\delta}(\kappa_{t_i}, \tau_i, \sigma) B^{2+\delta}(\kappa_{t_i}, \tau_i, \sigma)^{2+\delta} Z^{2+\delta}(\kappa_{t_i}, \tau_i) \\
&= \frac{1}{h_{1,n}^{1+\delta} h_{2,n}^{1+\delta}} E_t \left[\int A^{2+\delta}(\kappa_t - h_n u, \tau - h_n v, \sigma) \right. \\
&\quad \times B^{2+\delta}(\kappa_t - h_n u, \tau - h_n u, \sigma) \\
&\quad \left. \times Z^{2+\delta}(\kappa_t - h_{1,n} u, \tau - h_{2,n} v) \, du \, dv \right] \\
&= \frac{f_t(\kappa_t, \tau) \int K_{(1)}^{2+\delta}(u) \, du \int K_{(2)}^{2+\delta}(v) \, dv}{h_{1,n}^{1+\delta} h_{2,n}^{1+\delta}} \\
&\quad \times E_t \left[A^{2+\delta}(\kappa_t, \tau, \sigma) B^{2+\delta}(\kappa_t, \tau, \sigma) w^{2+\delta}(\kappa_t) \right] \\
&\quad + o\left(\frac{1}{h_{1,n}^{1+\delta} h_{2,n}^{1+\delta}} \right). \quad (C.17)
\end{aligned}$$

Similarly, we get:

$$\begin{aligned}
E_t u_{ni}^2 &= \frac{f_t(\kappa_t, \tau) \int K_{(1)}^2(u) \, du \int K_{(2)}^2(v) \, dv}{h_{1,n} h_{2,n}} \\
&\quad \times E_t \{ A^2(\kappa_t, \tau, \sigma) B^2(\kappa, \tau, \sigma) w^2(\kappa_t) \} \\
&\quad + o\left(\frac{1}{h_{1,n} h_{2,n}} \right). \quad (C.18)
\end{aligned}$$

Equations (C.17) and (C.18) together prove

$$\frac{\sum_{i=1}^n E_t|u_{ni}|^{2+\delta}}{\left(\sum_{i=1}^n E_t|u_{ni}|^2 \right)^{\frac{2+\delta}{2}}} = \mathcal{O}((nh_{1,n}h_{2,n})^{-\frac{\delta}{2}}) = o(1) \quad (C.19)$$

as $nh_{1,n}h_{2,n} \to 0$.

Applying the Liapounov central limit theorem, we get

$$\sqrt{nh_{1,n}h_{2,n}} \, U_n(\sigma) \xrightarrow{\mathcal{L}} N\left(0, f_t(\kappa_t, \tau) \nu^2\right), \quad (C.20)$$

where

$$\nu^2 \stackrel{\text{def}}{=} E_t \{ A^2(\kappa_t, \tau, \sigma) B^2(\kappa_t, \tau, \sigma) w^2(\kappa_t) \} \int K_{(1)}^2(u) K_{(2)}^2(v) \, du \, dv. \quad (C.21)$$

By (C.15) and (C.20), asymptotic normality is proved.

References

Airoldi, J.-P. and Flury, B. D. (1988). An application of common principal component analysis to cranial morphometry of microtus californicus and m. ochrogaster (mammalia, rodentia), *Journal of Zoology, Lond.* **216**: 21–36.

Aït-Sahalia, Y. (1996). Nonparametric pricing of interest rate derivative securities, *Econometrica* **64**: 527–560.

Aït-Sahalia, Y. and Duarte, J. (2003). Nonparametric option pricing under shape restrictions, *Journal of Econometrics* **116**: 9–47.

Aït-Sahalia, Y. and Lo, A. (1998). Nonparametric estimation of state-price densities implicit in financial asset prices, *Journal of Finance* **53**: 499–548.

Aït-Sahalia, Y., Bickel, P. J. and Stoker, T. M. (2001a). Goodness-of-fit tests for regression using kernel methods, *Journal of Econometrics* **105**: 363–412.

Aït-Sahalia, Y., Wang, Y. and Yared, F. (2001b). Do options markets correctly price the probabilities of movement of the underlying asset?, *Journal of Econometrics* **102**: 67–110.

Akaike, H. (1973). Information theory and an extension of the maximum likelihood principle, *2nd International Symposium on Information Theory*, Akademiai Kiado, Budapest.

Alexander, C. (2001a). *Market Models*, John Wiley & Sons, New York.

Alexander, C. (2001b). Principles of the skew, *RISK* **14**(1): S29–S32.

Alexander, C. and Lvov, D. (2003). Statistical properties of forward rates, *Working paper*, ISMA Centre, University of Reading.

Alexander, C. and Nogueira, L. M. (2004). Hedging with stochastic local volatility, *Discussion Papers in Finance 2004-11*, ISMA Centre, University of Reading.

Alexander, C., Brintalos, G. and Nogueira, L. (2003). Short and long term smile effects: The binomial normal mixture diffusion model, *Working paper*, ISMA Centre, University of Reading.

Amerio, E., Fusai, G. and Vulcano, A. (2003). Pricing of implied volatility derivatives, *FORC Preprint 2003/126*, University of Warwick.

Amin, K. I. and Ng, V. K. (1997). Inferring future volatility from the information in implied volatility in eurodollar options: A new approach, *Review of Financial Studies* **10**(2): 333–367.

References

Andersen, L. B. G. and Brotherton-Ratcliffe, R. (1997). The equity option volatility smile: An implicit finite-difference approach, *Journal of Computational Finance* **1**(2): 5–37.

Andersen, L. B. G., Andreasen, J. and Eliezer, D. (2002). Static replication of barrier options: Some general results, *Journal of Computational Finance* **5**(4): 1–25.

Andersen, T. G., Bollerslev, T., Diebold, F. X. and Labys, P. (2003). Modelling and forecasting realized volatility, *Econometrica* **71**: 579–625.

Anderson, T. W. (1963). Asymptotic theory for principal component analysis, *Annals of Mathematical Statistics* **34**: 122–148.

Ané, T. and Geman, H. (1999). Stochastic volatility and transaction time: An activity-based volatility estimator, *Journal of Risk* **2**(1): 57–69.

Avellaneda, M., Boyer-Olson, D., Busca, J. and Friz, P. (2002). Reconstructing volatility, *RISK* **15**(10): 91–95.

Avellaneda, M., Friedman, C., Holmes, R. and Samperi, D. (1997). Calibrating volatility surfaces via relative entropy minimization, *Applied Mathematical Finance* **4**: 37–64.

Ayache, E., Henrotte, P., Nassar, S. and Wang, X. (2004). Can anyone solve the smile problem?, *Wilmott magazine* (Jan.): 78–96.

Bajeux, I. and Rochet, J. C. (1992). Dynamic spanning: Are options an appropriate instrument?, *Mathematical Finance* **6**: 1–16.

Bakshi, G. and Kapadia, N. (2003). Delta-hedged gains and the negative market volatility risk premium, *Review of Financial Studies* **16**(2): 527–566.

Bakshi, G., Cao, C. and Chen, Z. (1997). Empirical performance of alternative option pricing models, *Journal of Finance* **52**(5): 2003–2049.

Bakshi, G., Cao, C. and Chen, Z. (2000). Do call and underlying prices always move in the same direction?, *Review of Financial Studies* **13**(3): 549–584.

Bakshi, G., Kapadia, N. and Madan, D. (2003). Stock return characteristics, skew laws, and the differential pricing of individual equity options, *Review of Financial Studies* **16**(1): 101–143.

Ball, C. and Roma, A. (1994). Stochastic volatility option pricing, *Journal of Financial and Quantitative Analysis* **29**(4): 589–607.

Balland, P. (2002). Deterministic implied volatility models, *Quantitative Finance* **2**: 31–44.

Barle, S. and Cakici, N. (1998). How to grow a smiling tree, *The Journal of Financial Engineering* **7**: 127–146.

Barndorff-Nielsen, O. E. (1997). Normal inverse Gaussian distributions and stochastic volatility modelling, *Scandinavian Journal of Statistics* **24**: 1–13.

Bates, D. S. (1996). Jumps and stochastic volatility: Exchange rate processes implicit in deutsche mark options, *Review of Financial Studies* **9**: 69–107.

Bates, D. S. (2000). Post-'87 crash fears in the S&P 500 futures option market, *Journal of Econometrics* **94**(1-2): 181–238.

Beaglehole, D. and Chebanier, A. (2002). Mean-reverting smiles, *RISK* **15**(4): 95–98.

Beckers, S. (1981). Standard deviations implied in option prices as predictors of future stock price variability, *Journal of Banking and Finance* **5**: 363–382.

Benko, M. and Härdle, W. (2004). Common functional implied volatility analysis, *in* P. Čížek, W. Härdle and R. Weron (eds), *Statistical Tools in Finance*, Springer-Verlag, Berlin, Heidelberg. Forthcoming.

Berestycki, H., Busca, J. and Florent, I. (2002). Asymptotics and calibration of local volatility models, *Quantitative Finance* **2**: 61–69.

Besse, P. (1991). Approximation spline de l'analyse en composantes principales d'une variable aléatoire hilbertienne, *Annales de la Faculté des Sciences de Toulouse* **12**: 329–346.

Björk, T. (1998). *Arbitrage Theory in Continuous Time*, Oxford University Press, Oxford.

Black, F. (1976). Studies of stock price volatility changes, *Proceedings of the 1976 Meetings of the American Statistical Association* pp. 177–181.

Black, F. (1992). Living up to the model, *in* P. Field and R. Jaycobs (eds), *From Black-Scholes to Black Holes: New Frontiers in Option Pricing*, Risk Magazine Ltd, London, pp. 17–20.

Black, F. and Scholes, M. (1973). The pricing of options and corporate liabilities, *Journal of Political Economy* **81**: 637–654.

Blaskowitz, O., Härdle, W. and Schmidt, P. (2004). Skewness and kurtosis trades, *in* S. T. Rachev (ed.), *Computational and Numerical Methods in Finance*, Birkhäuser.

Bliss, R. (1997). Movements in the term structure of interest rates, *Economic Review Q IV*, Federal Reserve Bank of Atlanta.

Bluman, G. (1980). On the transformation of diffusion processes into Wiener processes, *SIAM Journal on Applied Mathematics* **39**(2): 238–247.

Bodurtha, J. N. (2000). A linearization-based solution to the ill-posed local volatility estimation problem, *Working paper*, Georgetown University.

Bodurtha, J. N. and Jermakyan, M. (1999). Nonparametric estimation of an implied volatility surface, *Journal of Computational Finance* **2**(4): 29–60.

Bollen, N. and Whaley, R. E. (2003). Does net buying pressure affect the shape of the implied volatility functions?, *Working paper*.

Borak, S., Fengler, M. R., Härdle, W. and Mammen, E. (2005). Semiparametric state space factor models, *CASE Discussion Paper*, Humboldt-Universität zu Berlin.

Bouchouev, I. and Isakov, V. (1999). Uniqueness, stability and numerical methods for the inverse problem that arises in financial markets, *Inverse Problems* **15**: R95–R116.

Brace, A., Goldys, B., Klebaner, F. and Womersley, R. (2001). Market model of stochastic implied volatility with application to the BGM model, *Working paper*, Department of Statistics, University of New South Wales, Sydney.

Branger, N. and Schlag, C. (2004). Why is the index smile so steep?, *Review of Finance* **8**: 109–127.

Breeden, D. and Litzenberger, R. (1978). Price of state-contingent claims implicit in options prices, *Journal of Business* **51**: 621–651.

Breidt, F. J., Crato, N. and de Lima, P. (1998). The detection and estimation of long memory in stochastic volatility, *Journal of Econometrics* **83**: 325–348.

Brigo, D. and Mercurio, F. (2001). Displaced and mixture diffusions for analytically-tractable smile models, *in* H. German, D. B. Madan, S. R. Pliska and A. C. F. Vorst (eds), *Mathematical Finance Bachelier Congress 2000*, Springer-Verlag, Berlin, Heidelberg.

Brigo, D. and Mercurio, F. (2002). Log-normal-mixture dynamics and calibration to market volatility smiles, *International Journal of Theoretical and Applied Finance* **5**(4): 427–446.

Brigo, D., Mercurio, F. and Sartorelli, G. (2002). Alternative asset price dynamics and volatility smile, *Banca IMI report*.

Britten-Jones, M. and Neuberger, A. J. (2000). Option prices, implied price processes, and stochastic volatility, *Journal of Finance* **55**(2): 839–866.

Broadie, M., Cvitanić, J. and Soner, H. M. (1998). Optimal replication of contingent claims under portfolio constraints, *Review of Financial Studies* **11**(1): 59–79.

Broadie, M., Detemple, J., Ghysels, E. and Torrès, O. (2000). American options with stochastics dividends and volatility: A nonparametric investigation, *Journal of Econometrics* **94**: 53–92.

Brown, G. and Randall, C. (1999). If the skew fits, *RISK* **12**(4): 62–65.

Brunner, B. and Hafner, R. (2003). Arbitrage-free estimation of the risk-neutral density from the implied volatility smile, *Journal of Computational Finance* **7**(1): 75–106.

Cai, Z., Fan, J. and Yao, Q. (2000). Functional-coefficient regression models for nonlinear time series, *Journal of the American Statistical Association* **95**: 941–956.

Canina, L. and Figlewski, S. (1993). The informational content of implied volatility, *Review of Financial Studies* **6**: 659–681.

Carr, P. and Madan, D. (1998). Towards a theory of volatility trading, *in* R. Jarrow (ed.), *Volatility*, Risk Publications, pp. 417–427.

Carr, P., Ellis, K. and Gupta, V. (1998). Static hedging of exotic options, *Journal of Finance* **53**(3): 1165–1190.

Chiras, D. P. and Manaster, S. (1978). The information content of option prices and a test for market efficiency, *Journal of Financial Economics* **6**: 213–234.

Christensen, B. and Prabhala, N. (1998). The relation between implied and realized volatility, *Journal of Financial Economics* **50**: 125–150.

Čížek, P., Härdle, W. and Weron, R. (2004). *Statistical Tools in Finance*, Springer-Verlag, Berlin, Heidelberg. Forthcoming.

Coleman, T. F., Kim, Y., Li, Y. and Verma, A. (2001). Dynamic hedging with a deterministic local volatility function model, *Journal of Risk* **4**(1): 63–89.

Coleman, T. F., Li, Y. and Verma, A. (1999). Reconstructing the unknown local volatility function, *Journal of Computational Finance* **2**(3): 77–102.

Connor, G. and Linton, O. (2000). Semiparametric estimation of a characteristic-based factor model of stock returns, *Technical report*, LSE, London.

Cont, R. (1999). Beyond implied volatility: Extracting information from option prices, *in* I. Kondor and J. Kertesz (eds), *Econophysics: An Emerging Science*, Kluwer Academic Publishers, Dordrecht.

Cont, R. and da Fonseca, J. (2002). The dynamics of implied volatility surfaces, *Quantitative Finance* **2**(1): 45–60.

Cont, R. and Tankov, P. (2003). Calibration of jump-diffusion option pricing models: A robust non-parametric approach, *Journal of Computational Finance*. Forthcoming.

Cont, R. and Tankov, P. (2004). *Financial Modelling with Jump Processes*, Chapman & Hall, CRC Press, London.

Cont, R., da Fonseca, J. and Durrleman, V. (2002). Stochastic models of implied volatility surfaces, *Economic Notes* **31**(2): 361–377.

Corrozet, G. (1543). *Hecaton-GRAPHIE. C'est à dire les descriptions de cent figures & hystoires, contenants plusieurs appophthegmes, prouerbes, sentences & dictz tant des anciens, que des modernes. Le tout reueu par son autheur. Auecq'Priuilege. A Paris chez Denys Ianot Imprimeur & Libraire.*

Cox, J. E. and Ross, S. A. (1976). The valuation of options for alternative stochastic processes, *Journal of Financial Economics* **76**: 145–166.
Cox, J. E., Ross, S. A. and Rubinstein, M. (1979). Option pricing: A simplified approach, *Journal of Financial Economics* **7**: 229–263.
Crépey, S. (2004). Delta-hedging vega risk?, *Technical report*, Université d'Évry, France.
Daglish, T. (2003). Pricing and hedging comparison for index options, *Journal of Financial Econometrics* **1**(3): 327–364.
Daglish, T., Hull, J. C. and Suo, W. (2003). Volatility surfaces: Theory, rules of thumb, and empirical evidence, *Working paper*, J. L. Rotman School of Management, University of Toronto.
Das, S. and Sundaram, R. (1999). Of smiles and smirks: A term-structure perspective, *Journal of Financial and Quantitative Analysis* **34**(2): 211–240.
Dauxois, J., Pousse, A. and Romain, Y. (1982). Asymptotic theory for the principal component analysis of a vector random function: Some applications to statistical inference, *Journal of Multivariate Analysis* **12**: 136–154.
Dempster, M. A. H. and Richards, D. G. (2000). Pricing American options fitting the smile, *Mathematical Finance* **10**(2): 157–177.
Derman, E. (1999). Regimes of volatility, *RISK* **12**(4): 55–59.
Derman, E. and Kani, I. (1994a). Riding on a smile, *RISK* **7**(2): 32–39.
Derman, E. and Kani, I. (1994b). The volatility smile and its implied tree, *Quantitative strategies research notes*, Goldman Sachs.
Derman, E. and Kani, I. (1998). Stochastic implied trees: Arbitrage pricing with stochastic term and strike structure of volatility, *International Journal of Theoretical and Applied Finance* **1**(1): 61–110.
Derman, E., Ergener, D. and Kani, I. (1995). Static options replication, *Journal of Derivatives* **2**(4): 78–95.
Derman, E., Kani, I. and Chriss, N. (1996a). Implied trinomial trees of the volatility smile, *Journal of Derivatives* **3**(4): 7–22.
Derman, E., Kani, I. and Kamal, M. (1997). Trading and hedging local volatility, *Journal of Financial Engineering* **6**(3): 1233–1268.
Derman, E., Kani, I. and Zou, J. Z. (1996b). The local volatility surface: Unlocking the information in index option prices, *Financial Analysts Journal* **7-8**: 25–36.
Deutsche Börse (2002). *Leitfaden zu den Aktienindizes der Deutschen Börse*, 4.3 edn, Deutsche Börse AG, 60284 Frankfurt am Main.
Duffie, D. (2001). *Dynamic Asset Pricing Theory*, 3rd edn, Princeton University Press, Princeton.
Dumas, B., Fleming, J. and Whaley, R. E. (1998). Implied volatility functions: Empirical tests, *Journal of Finance* **80**(6): 2059–2106.
Dupire, B. (1994). Pricing with a smile, *RISK* **7**(1): 18–20.
Eberlein, E. and Keller, U. (1995). Hyperbolic distributions in finance, *Bernoulli* **1**: 281–299.
Eberlein, E. and Prause, K. (2002). The generalized hyperbolic model: Financial derivatives and risk measures, *in* H. Geman, D. Madan, S. Pliska and T. Vorst (eds), *Mathematical Finance – Bachelier Congress 2000*, Springer-Verlag, Berlin, Heidelberg, pp. 245–267.
Ederington, L. and Guan, W. (2002). Why are those options smiling?, *Journal of Derivatives* **10**(2): 9–34.

Efromovich, S. (1999). *Nonparametric Curve Estimation*, Springer-Verlag, Berlin, Heidelberg.

Engle, R. (1982). Autoregressive conditional heteroscedasticity with estimates of the variance of United Kingdom inflation, *Econometrica* **50**(4): 987–1007.

Engle, R. (2002). Dynamical conditional correlation: A simple class of multivariate generalized autoregressive conditional heteroscedastic models, *Journal of Business and Economic Statistics* **20**(3): 339–350. Forthcoming.

Engle, R. and Rosenberg, J. (2000). Testing the volatility term structure using option hedging criteria, *Journal of Derivatives* **8**(1): 10–28.

Evans, M., Hastings, N. and Peacock, B. (2000). *Statistical Distributions*, 3rd edn, John Wiley & Sons, New York.

Fahlenbrach, R. and Strobl, G. (2002). Is the volatility constrained to smile? An empiricial investigation of option pricing models under portfolio constraints, *Working paper*, University of Pennsylvania.

Fan, J. (1992). Design adaptive nonparametric regression, *Journal of the American Statistical Association* **87**: 998–1004.

Fan, J. (1993). Local linear regression smoothers and their minimax efficiencies, *Journal of the American Statistical Association* **21**: 196–216.

Fan, J. and Gijbels, I. (1992). Variable bandwidth and local linear regression smoothers, *Annals of Statistics* **21**: 196–216.

Fan, J., Yao, Q. and Cai, Z. (2003). Adaptive varying-coefficient linear models, *J. Roy. Statist. Soc. B.* **65**: 57–80.

Fengler, M. R. (2002). The phenomenology of implied volatility surfaces, *Master thesis*. Department of Business and Economics, Humboldt-Universität zu Berlin.

Fengler, M. R. and Herwartz, H. (2002). Multivariate volatility models, *in* W. Härdle, T. Kleinow and G. Stahl (eds), *Applied Quantitative Finance*, Springer-Verlag, Berlin, Heidelberg.

Fengler, M. R. and Schwendner, P. (2004). Quoting multiasset equity options in the presence of errors from estimating correlations, *Journal of Derivatives* **11**(4): 43–54.

Fengler, M. R. and Wang, Q. (2003). Fitting the smile revisited: A least squares kernel estimator for the implied volatility surface, *SfB 373 Discussion Paper 2003-25*, Humboldt-Universität zu Berlin.

Fengler, M. R. and Winter, J. (2004). Price variability and price dispersion in a stable monetary environment: Evidence from Germany, *Managerial and Decision Economics*. Special Issue on *Price Flexibility: Theories and Evidence*, D. Levy (ed.), forthcoming.

Fengler, M. R., Härdle, W. and Mammen, E. (2003a). A dynamic semiparametric factor model for implied volatility string dynamics, *Discussion paper*, SfB 373, Humboldt-Universität zu Berlin.

Fengler, M. R., Härdle, W. and Schmidt, P. (2002a). The analysis of implied volatilities, *in* W. Härdle, T. Kleinow and G. Stahl (eds), *Applied Quantitative Finance*, Springer-Verlag, Berlin, Heidelberg.

Fengler, M. R., Härdle, W. and Schmidt, P. (2002b). Common factors governing VDAX movements and the maximum loss, *Journal of Financial Markets and Portfolio Management* **16**(1): 16–29.

Fengler, M. R., Härdle, W. and Villa, C. (2003b). The dynamics of implied volatilities: A common principle components approach, *Review of Derivatives Research* **6**: 179–202.

Figlewski, S. (1989). What does an option pricing model tell us about option prices?, *Financial Analysts Journal* **45**: 12–15.

Flury, B. (1988). *Common Principal Components and Related Multivariate Models*, Wiley Series in Probability and Mathematical Statistics, John Wiley & Son, New York.

Flury, B. and Gautschi, W. (1986). An algorithm for simultaneous orthogonal transformations of several positive definite matrices to nearly diagonal form, *Journal on Scientific and Statistical Computing* **7**: 169–184.

Föllmer, H. and Schied, A. (2002). *Stochastic Finance: An Introduction in Discrete Time*, Wiley Series in Probability and Mathematical Statistics, Walter de Gruyter, Berlin, New York.

Föllmer, H. and Schweizer, M. (1990). Hedging of contingent claims under incomplete information, *in* M. H. A. Davis and R. J. Elliott (eds), *Applied Stochastical Analysis*, Vol. 5 of *Stochastics Monographs*, Gordon and Breach, New York, pp. 389–414.

Föllmer, H. and Sondermann, D. (1986). Hedging of non-redundant contingent claims, *in* W. Hildenbrand and A. Mas-Colell (eds), *Contributions to Mathematical Economics in Honor of Gérard Debreu*, North-Holland, Amsterdam, pp. 206–223.

Fouque, J.-P., Papanicolaou, G. and Sircar, K. R. (2000). *Derivatives in Financial Markets with Stochastic Volatility*, Cambridge University Press, Cambridge.

Franke, J., Härdle, W. and Hafner, C. (2004). *Introduction to the Statistics of Financial Markets*, Springer-Verlag, Berlin, Heidelberg. Forthcoming.

Frey, R. (1996). Derivative asset analysis in models with level-dependent and stochastic volatility, *CWI Quarterly* **10**(1): 1–34.

Frey, R. and Patie, P. (2002). Risk management for derivatives in illiquid markets: A simulation study, *in* K. Sandmann and P. Schönbucher (eds), *Advances in Finance and Stochastics*, Springer-Verlag, Berlin, Heidelberg.

Gatheral, J. (1999). The volatility skew: Arbitrage constraints and asymptotic behavior, *Technical report*, Merill Lynch.

Ghysels, E. and Ng, S. (1989). A semiparametric factor model of interest rates and tests of the affine term structure, *Review of Economics and Statistics* **80**: 535–548.

Glosten, L., Jagannathan, R. and Runkle, D. (1993). Relationship between the expected value and the volatility of the nominal excess return on stocks, *Journal of Finance* **48**: 1779–1801.

Golub, B. and Tilman, L. M. (1997). Measuring yield curve risk using principal component analysis, value at risk, and key rate durations, *Journal of Portfolio Management* **23**(4): 72–84.

Gouriéroux, C. and Jasiak, J. (2001). Dynamic factor models, *Econometrics Review* **20**(4): 385–424.

Gouriéroux, C., Monfort, A. and Tenreiro, C. (1994). Nonparametric diagnostics for structural models, *Document de travail 9405*, CREST, Paris.

Gouriéroux, C., Monfort, A. and Tenreiro, C. (1995). Kernel M-estimators and functional residual plots, *Document de travail 9546*, CREST, Paris.

Gouriéroux, C., Scaillet, O. and Szafarz, A. (1997). *Econométrie de la finance*, Economica, Paris.

Grossman, S. and Zhou, Z. (1996). Equilibrium analysis of portfolio insurance, *Journal of Finance* **51**(4): 1379–1403.

214 References

Hafner, R. and Wallmeier, M. (2001). The dynamics of DAX implied volatilities, *International Quarterly Journal of Finance* **1**(1): 1–27.

Hagan, P. and Woodward, D. (1999). Equivalent Black volatilities, *Applied Mathematical Finance* **6**: 147–157.

Hagan, P., Kumar, D., Lesniewski, A. and Woodward, D. (2002). Managing smile risk, *Wilmott magazine* **1**: 84–108.

Härdle, W. (1990). *Applied Nonparametric Regression*, Cambridge University Press, Cambridge, UK.

Härdle, W. and Hafner, C. (2000). Discrete time option pricing with flexible volatility estimation, *Finance and Stochastics* **4**(2): 189–207.

Härdle, W. and Hlávka, Z. (2004). Dynamics of state price densities, *CASE Discussion Paper*, Humboldt-Universität zu Berlin.

Härdle, W. and Simar, L. (2003). *Applied Multivariate Statistical Analysis*, Springer-Verlag, Berlin, Heidelberg.

Härdle, W. and Yatchew, A. (2003). Dynamic state price density estimation using constrained least squares and the bootstrap, *Journal of Econometrics*. Forthcoming.

Härdle, W. and Zheng, J. (2002). How precise are price distributions predicted by implied binomial trees?, *in* W. Härdle, T. Kleinow and G. Stahl (eds), *Applied Quantitative Finance*, Springer-Verlag, Berlin, Heidelberg.

Härdle, W., Herwartz, H. and Spokoiny, V. (2003). Time inhomogeneous multiple volatility modelling, *Journal Financial Econometrics* **1**(2): 55–95.

Härdle, W., Hlávka, Z. and Klinke, S. (2000a). *XploRe – Application Guide*, Springer-Verlag, Berlin, Heidelberg.

Härdle, W., Kleinow, T. and Stahl, G. (2002). *Applied Quantitative Finance*, Springer-Verlag, Berlin, Heidelberg.

Härdle, W., Klinke, S. and Müller, M. (2000b). *Xplore – Learning Guide*, Springer-Verlag, Berlin, Heidelberg.

Härdle, W., Müller, M., Sperlich, S. and Werwatz, A. (2004). *Nonparametric and Semiparametric Models*, Springer-Verlag, Berlin, Heidelberg.

Harper, J. (1994). Reducing parabolic partial differential equations to canonical form, *European Journal of Applied Mathematics* **5**: 159–165.

Harrison, J. and Kreps, D. (1979). Martingales and arbitrage in multiperiod securities markets, *Journal of Economic Theory* **20**: 381–408.

Harvey, C. R. and Whaley, R. E. (1991). S&P 100 index option volatility, *Journal of Finance* **46**(4): 1151–1561.

Harvey, C. R. and Whaley, R. E. (1992). Market volatility prediction and the efficiency of the S&P 100 index option market, *Journal of Financial Economics* **31**: 43–73.

Hastie, T. and Tibshirani, R. (1990). *Generalized additive models*, Chapman and Hall, London.

Heath, D., Jarrow, R. A. and Morton, A. (1992). Bond pricing and the term structure of interest rates: A new methodology for contingent claims valuation, *Econometrica* **60**: 77–105.

Henkel, A. and Schöne, A. (1996). *Emblemata. Handbuch zur Sinnbildkunst des XVI. und XVII. Jahrhunderts*, Verlag J. B. Metzler, Stuttgart, Weimar.

Hentschel, L. (2003). Errors in implied volatility estimation, *Journal of Financial and Quantitative Analysis* **38**: 779–810.

Heston, S. (1993). A closed-form solution for options with stochastic volatility with applications to bond and currency options, *Review of Financial Studies* **6**: 327–343.

Heynen, R. (1994). An empirical investigation of observed smile patterns, *Review of Futures Markets* **13**: 317–353.

Hlávka, Z. (2003). Constrained estimation of state price densities, *Discussion Paper 2003-22*, SfB 373, Humboldt-Universität zu Berlin.

Hormander, L. (1990). *The Analysis of Linear Partial Differential Operators I: Distribution Theory and Fourier Analysis*, 2nd edn, Springer-Verlag, Berlin, Heidelberg.

Horowitz, J. (1998). *Semiparametric Methods in Econometrics*, number 131 in Lecture Notes in Statistics, Springer-Verlag, Berlin, Heidelberg.

Horowitz, J., Klemela, J. and Mammen, E. (2002). Optimal estimation in additive models, *Preprint*.

Hotelling, H. (1933). Analysis of a complex of statistical variables into principal components, *Journal of Educational Psychology* **24**: 417–441.

Hull, J. (2002). *Options, Futures, and Other Derivatives*, Prentice Hall, New Jersey, USA.

Hull, J. and White, A. (1987). The pricing of options on assets with stochastic volatilities, *Journal of Finance* **42**: 281–300.

Huynh, K., Kervalla, P. and Zheng, J. (2002). Estimating state price densities with nonparametric regression, *in* W. Härdle, T. Kleinow and G. Stahl (eds), *Applied Quantitative Finance*, Springer-Verlag, Berlin, Heidelberg.

Ingersoll, J. E. (1997). Valuing foreign exchange rate derivatives with a bounded exchange rate process, *Review of Derivatives Research* **1**: 159–181.

Jackson, N., Süli, E. and Howison, S. (1998). Computation of deterministic volatility surfaces, *Journal of Computational Finance* **2**(2): 5–32.

Jackwerth, J. C. (1997). Generalized binomial trees, *Journal of Derivatives* **5**: 7–17.

Jackwerth, J. C. (1999). Option-implied risk-neutral distributions and implied binomial trees: A literature review, *Journal of Derivatives* **7**(2): 66–82.

Jackwerth, J. C. and Rubinstein, M. (2001). Recovering stochastic processes from option prices, *Working paper*, Universität Konstanz.

Jamshidian, F. (1993). Options and futures evaluation with deterministic volatilities, *Mathematical Finance* **3**(2): 149–159.

Jamshidian, F. and Zhu, Y. (1997). Scenario simulation: Theory and methodology, *Finance and Stochastics* **1**: 43–67.

Jarrow, R. A. and O'Hara, M. (1989). Primes and scores: An essay on market imperfections, *Journal of Finance* **44**: 1265–1287.

Jiang, L. and Tao, Y. (2001). Identifying the volatility of the underlying assets from option prices, *Inverse Problems* **17**: 137–155.

Jiang, L., Chen, Q., Wang, L. and Zhang, J. E. (2003). A new well-posed algorithm to recover implied local volatility, *Quantitative Finance* **3**: 451–457.

Johnson, R. A. and Wichern, D. W. (1998). *Applied Multivariate Statistical Analysis*, 4 edn, Prentice-Hall, Englewood Cliffs, N.J.

Jorion, P. (1988). On jump processes in the foreign exchange and stock markets, *Review of Financial Studies* **1**(4): 427–445.

Jorion, P. (1995). Predicting volatility in the foreign exchange market, *Journal of Finance* **50**(2): 507–528.

Joshi, M. S. (2003). *The Concepts and Practice of Mathematical Finance*, Cambridge University Press, Cambridge.

Karatzas, I. (1997). *Lectures on the Mathematics of Finance*, Vol. 8 of *CRM Monograph Series*, American Mathematical Society, Providence, Rhode Island.

Karatzas, I. and Shreve, S. E. (1991). *Brownian Motion and Stochastic Calculus*, Springer-Verlag, Berlin, Heidelberg.

Khatri, C. G. (1980). Quadratic forms in normal variables, *in* P. R. Krishnaiah (ed.), *Handbook of Statistics*, Vol. I, North-Holland Publishing Company, Amsterdam, New York, Oxford, Tokyo, pp. 443–469.

Kruse, S. (2003). On the pricing of forward starting options under stochastic volatility, *Berichte des Fraunhofer ITWM 53(2003)*, Fraunhofer Institut Techno- und Wirtschaftsmathematik, Kaiserslautern.

Küchler, U., Neumann, K., Sørensen, M. and Streller, A. (1999). Stock returns and hyperbolic distributions, *Mathematical and Computer Modelling* **29**: 1–15.

Lagnado, R. and Osher, S. (1997). A technique for calibrating derivative security pricing models: Numerical solution of an inverse problem, *Journal of Computational Finance* **1**(1): 13–25.

Lamoureux, C. G. and Lastrapes, W. D. (1993). Forecasting stock-return variance: Toward an understanding of stochastic implied volatilities, *Review of Financial Studies* **6**(2): 293–326.

Latané, H. A. and Rendelman, J. (1976). Standard deviations of stock price ratios implied in option prices, *Journal of Finance* **31**: 369–381.

Ledoit, O. and Santa-Clara, P. (1998). Relative option pricing with stochastic volatility, *Working paper*, UCLA, Los Angeles, USA.

Lee, P., Wang, L. and Karim, A. (2003). Index volatility surface via moment-matching techniques, *RISK* **16**(12): 85–89.

Lee, R. W. (2001). Implied and local volatilities under stochastic volatility, *International Journal of Theoretical and Applied Finance* **4**(1): 45–89.

Lee, R. W. (2002). Implied volatility: Statics, dynamics, and probabilistic interpretation, *Recent Advances in Applied Probability*. Forthcoming.

Lee, R. W. (2003). The moment formula for implied volatility at extreme strikes, *Mathematical Finance*. Forthcoming.

Lepski, O. and Spokoiny, V. (1997). Optimal pointwise adaptive methods in nonparametric estimation, *Annals of Statistics* **25**: 2512–2546.

Lewis, A. L. (2000). *Option Valuation under Stochastic Volatility*, Finance Press.

Lintner, J. (1965). The valuation of risky assets and the selection of risky investments in stock portfolios and capital budgets, *Review of Economics and Statistics* **47**: 13–37.

Linton, O., Mammen, E., Nielsen, J. and Tanggaard, C. (2001). Yield curve estimation by kernel smoothing, *Journal of Econometrics* **105**(1): 185–223.

Linton, O., Nguyen, T. and Jeffrey, A. (2003). Nonparametric estimation of single factor Heath-Jarrow-Morton term structure models and a test for path independence, *Technical report*, LSE, London.

Lipton, A. (2001). *Mathematical Methods For Foreign Exchange: A Financial Engineer's Approach*, World Scientific Publishing Company.

Manaster, S. and Koehler, G. (1982). The calculation of implied variances from the Black-and-Scholes model: A note, *Journal of Finance* **37**: 227–230.

Mardia, K. V., Kent, J. T. and Bibby, J. M. (1992). *Multivariate Analysis*, 8th edn, Academic Press, Academic Press Ltd., London.

Markowitz, H. (1959). *Portfolio Selection: Efficient Diversification of Investments*, John Wiley, New York.

Marron, J. S. and Härdle, W. (1986). Random approximations to an error criterion of nonparametric statistics, *Journal of Multivariate Analysis* **20**: 91–113.

Marron, J. S. and Nolan, D. (1988). Canonical kernels for density estimation, *Statistics and Probability Letters* **7**(3): 195–199.

McIntyre, M. L. (2001). Performance of Dupire's implied diffusion approach under sparse and incomplete data, *Journal of Computational Finance* **4**(4): 33–84.

Mercurio, D. (2004). *Adaptive estimation for financial time series*, PhD thesis, Humboldt-Universität zu Berlin, Berlin.

Mercurio, D. and Spokoiny, V. (2004). Statistical inference for time-inhomogeneous volatility models, *Annals of Statistics*. Forthcoming.

Merton, R. C. (1973). Theory of rational option pricing, *Bell Journal of Economics and Management Science* **4**(Spring): 141–183.

Merton, R. C. (1976). Option pricing when underlying stock returns are discontinuous, *Journal of Financial Economics* **3**: 125–144.

Meyer, P. A. (1976). *Un cours sur les intégrales stochastiques*, number 511 in *Lecture Notes in Mathematics*, Springer-Verlag, Berlin, Heidelberg.

Molgedey, L. and Galic, E. (2001). Extracting factors for interest rate scenarios, *European Physical Journal B* **20**(4): 517–522.

Musiela, M. and Rutkowski, M. (1997). *Martingale Methods for Financial Modelling*, Springer-Verlag, Berlin, Heidelberg.

Nadaraya, E. A. (1964). On estimating regression, *Theory of Probability and its Applications* **10**: 186–190.

Nagot, I. and Trommsdorff, R. (1999). The tree of knowledge, *RISK* **12**(8): 99–102.

Nelson, D. B. (1991). Conditional heteroskedasticity in asset returns: A new approach, *Econometrica* **59**: 347–370.

Nelson, D. B. and Ramaswamy, K. (1990). Simple binomial processes as diffusion approximations in financial models, *Review of Financial Studies* **3**(3): 393–430.

Niffikeer, C. L., Hewins, R. D. and Flavell, R. B. (2000). A synthetic factor approach to the estimation of value-at-risk of a portfolio of interest rate swaps, *Journal of Banking and Finance* **24**: 1903–1932.

Øksendal, B. (1998). *Stochastic Differential Equations*, 5th edn, Springer-Verlag, Berlin, Heidelberg.

Overhaus, M. (2002). Himalaya options, *RISK* **15**(3): 101–104.

Pagan, A. and Ullah, A. (1999). *Nonparametric Econometrics*, Cambridge University Press, Cambridge.

Pearson, K. (1901). On lines and planes of closest fit to systems of points in space, *Philosophical Magazine* **2**(6): 559–572.

Peña, I., Rubio, G. and Serna, G. (1999). Why do we smile? On the determinants of the implied volatility function, *Journal of Banking and Finance* **23**: 1151–1179.

Pérignon, C. and Villa, C. (2002). Component proponents, *RISK* **15**(9): 154–156.

Pérignon, C. and Villa, C. (2004). Component proponents II, *RISK* **17**(7): 77–79.

Pezzulli, S. and Silverman, B. W. (1993). Some properties of smoothed principal components analysis for functional data, *Computational Statistics* **8**: 1–13.

Pham, H. and Touzi, N. (1996). Intertemporal equilibrium risk premia in a stochastic volatility model, *Journal of Mathematical Finance* **6**: 215–236.

Pong, S., Shackleton, M., Taylor, S. and Xu, X. (2003). Forecasting currency volatility: A comparison of implied volatilities and AR(FI)MA models, *Journal of Banking and Finance*. Forthcoming.

Poon, S.-H. and Granger, C. W. J. (2003). Forecasting volatility in financial markets: A review, *Journal of Economic Literature* **41**: 478–539.

Press, W., Flannery, B., Teukolsky, S. and Vetterling, W. (1993). *Numerical Recipes in C: The Art of Scientific Computing*, 2nd edn, Cambridge University Press.

Quessette, R. (2002). New products, new risks, *RISK* **15**(3): 97–100.

Rady, S. (1997). Option pricing in the presence of natural boundaries and a quadratic diffusion term, *Finance and Stochastics* **1**: 331–344.

Ramsay, J. O. and Silverman, B. W. (1997). *Functional Data Analysis*, Springer-Verlag, Berlin, Heidelberg.

Randall, C. and Tavella, D. (2000). *Pricing Financial Instruments: The Finite Difference Method*, John Wiley & Sons, New York.

Rao, C. R. (1973). *Linear Statistical Inference and Its Applications*, 2nd edn, Wiley, New York.

Rebonato, R. (1998). *Interest-Rate Option Models: Understanding, Analyzing and Using Models for Exotic Interest-Rate Options*, Wiley Series in Financial Engineering, 2nd edn, John Wiley & Sons Ltd.

Rebonato, R. (1999). *Volatility and Correlation*, Wiley Series in Financial Engineering, John Wiley & Son Ltd.

Renault, E. and Touzi, N. (1996). Option hedging and implied volatilities in a stochastic volatility model, *Mathematical Finance* **6**(3): 279–302.

Riesz, F. and Nagy, B. (1956). *Functional Analysis*, Blackie, London.

Roll, R. (1984). A simple implicit measure of the effective bid-ask spread, *Journal of Finance* **39**: 1127–1139.

Rookley, C. (1997). Fully exploiting the information content of intra-day option quotes: Applications in option pricing and risk management, *Technical report*, Department of Finance, University of Arizona.

Rose, G. (2004). *Unternehmenssteuerrecht*, E. Schmidt Verlag.

Rosenberg, J. (2000). Implied volatility functions: A reprise, *Journal of Derivatives* **7**: 51–64.

Rossi, A. (2002). The Britten-Jones and Neuberger smile-consistent with stochastic volatility option pricing model: A further analysis, *International Journal of Theoretical and Applied Finance* **5**(1): 1–31.

Rubinstein, M. (1994). Implied binomial trees, *Journal of Finance* **49**: 771–818.

Ruppert, D. (1997). Empirical-bias bandwidths for local polynomial nonparametric regression and density estimation, *Journal of the American Statistical Association* **92**: 1049–1062.

Ruppert, D. and Wand, M. P. (1994). Multivariate locally weighted least squares regression, *Annals of Statistics* **22**(3): 1346–1370.

Schmalensee, R. and Trippi, R. R. (1978). Common stock volatility expectations implied by option premia, *Journal of Finance* **33**: 129–147.

Schönbucher, P. J. (1999). A market model for stochastic implied volatility, *Philosophical Transactions of the Royal Society* **357**(1758): 2071–2092.

Schoutens, W. (2003). *Lévy Processes in Finance*, John Wiley & Sons, New York.

Schwarz, G. (1978). Estimating the dimension of a model, *Annals of Statistics* **6**: 461–464.

Scott, L. (1987). Option pricing when the variance changes randomly: Theory, estimation, and an application, *Journal of Financial and Quantitative Analysis* **22**: 419–37.
Sharpe, W. (1964). Capital asset prices: A theory of market equilibrium under conditions of risk, *Journal of Finance* **19**: 425–442.
Shimko, D. (1993). Bounds on probability, *RISK* **6**(4): 33–37.
Shu, J. and Zhang, J. E. (2003). The relationship between implied and realized volatility of S&P 500 index, *Wilmott magazine* **Jan.**: 83–91.
Skiadopoulos, G. (2001). Volatility smile consistent option models: A survey, *International Journal of Theoretical and Applied Finance* **4**(3): 403–437.
Skiadopoulos, G., Hodges, S. and Clewlow, L. (1999). The dynamics of the S&P 500 implied volatility surface, *Review of Derivatives Research* **3**: 263–282.
Spokoiny, V. (1998). Estimation of a function with discontinuities via local polynomial fit with an adaptive window choice, *Annals of Statistics* **26**: 1356–1378.
Steele, J. M. (2000). *Stochastic Calculus and Financial Applications*, Springer-Verlag, Berlin, Heidelberg, New York.
Stein, E. M. and Stein, J. C. (1991). Stock price distributions with stochastic volatility: An analytic approach, *Review of Financial Studies* **4**: 727–752.
Stone, C. J. (1986). The dimensionality reduction principle for generalized additive models, *The Annals of Statistics* **14**: 592–606.
Tanaka, H. (1963). Note on continuous additive functionals of the 1-dimensional Brownian path, *Zeitschrift für Wahrscheinlichkeitstheorie* **1**: 251–257.
Taylor, S. J. (2000). Consequences for option pricing of a long memory in volatility, *Working paper*, Department of Accounting and Finance, Lancaster University, UK.
Tipke, K., Lang, J. and Seer, R. (2002). *Steuerrecht*, O. Schmidt Verlag, Köln.
Tompkins, R. (1999). Implied volatility surfaces: Uncovering regularities for options on financial futures, *Working paper*, Vienna University of Technology.
Tompkins, R. (2001). Stock index futures markets: Stochastic volatility models and smiles, *The Journal of Futures Markets* **21**(1): 43–78.
Tse, Y. and Tsui, A. (2002). A multivariate generalized autoregressive conditional heteroscedastic model with time-varying correlations, *Journal of Business and Economic Statistics* **20**(3): 351–362.
Vähämaa, S. (2004). Delta hedging with the smile, *Financial Markets and Portfolio Management* **18**(3): 241–255.
Watson, G. S. (1964). Smooth regression analysis, *Sankyhā, Series A* **26**: 359–372.
Weinberg, S. A. (2001). Interpreting the volatility smile: An examination of the informational content of option prices, *International Finance Discussion Papers 706*, Federal Reserve Board, Washington, D. C.
Whaley, R. (1982). Valuation of American call options on dividend-paying stocks: Empirical tests, *Journal of Financial Economics* **10**: 29–58.
Wilmott, P. (2001a). *Paul Wilmott on Quantitative Finance*, Vol. 1, John Wiley & Sons.
Wilmott, P. (2001b). *Paul Wilmott on Quantitative Finance*, Vol. 2, John Wiley & Sons.
Zakoian, J. M. (1994). Threshold heteroskedastic functions, *Journal of Economic Dynamics and Control* **18**: 931–955.
Zhu, Y. and Avellaneda, M. (1997). An E-ARCH model for the term-structure of implied volatility of FX options, *Applied Mathematical Finance* **4**: 81–100.

Zhu, Y. and Avellaneda, M. (1998). A risk-neutral stochastic volatility model, *International Journal of Theoretical and Applied Finance* **1**(2): 289–310.

Zühlsdorff, C. (2002). The pricing of derivatives on assets with quadratic volatility, *Working Paper B-451*, Bonn SfB 303.

Index

Akaike information criterion 106, 109, 112, 131, 168
arbitrage 11
at-the-money 20
average squared error 105

bandwidth choice 104–112
Barle Cakici implied tree 69
Black Scholes formula 19, 38, 48, 56, 115, 190
Black Scholes model 9–10
  call option 14
  generalized PDE 35, 50
  partial differential equation 12
bond, riskless 10
Brigo Mercurio model 85
Britten-Jones Neuberger implied tree 81
Brownian motion
  definition 196
  geometric 9

call option 10
  Black Scholes formula 14
common principal component models 128–149
  asymptotic distribution of eigenvalues 134
  asymptotic distribution of eigenvectors 134
  hierarchy of models 131
  likeklihood function 133
  model selection 138–139
  motivation 129

  of implied volatility surface 139–145
  partial 131
  proportional model 130
  stability analysis 145–149
  stability tests 134–137
  time series models 149–154
constant elasticity of variance model 84
contingent claim 10
counterparty 10
covariation process 196
cross validation 105, 106, 112

delta 15, 26, 40, 88
  model consistent 89
  recalibration 90
  sticky-moneyness 89
  sticky-strike 89
  vega correction 40, 89
delta hedging 15, 35, 40–42, 44, 88–90
delta-sigma hedging 39
derivative 10
derivatives estimation 57,
  → nonparametric regression
Derman Kani Criss implied tree 77
Derman Kani implied tree 69
Derman Kani stochastic implied tree 80
difference dividend 191, 192
dimension reduction
  → common principle component models, functional principle component analysis, semiparametric factor models

Index

Dupire formula 51, 53, 55
  discrete-time version 82
  implied volatility counterpart 56, 57, 112

exercise price 10
expectation
  K-strike, T-maturity forward
    risk-adjusted 51, 64–66
  risk neutral 12, 38, 49

filtration 195
Fokker-Planck equation 54, 199
forward price 20, 24, 191
functional data analysis 155–160
functional principle component analysis
  computation 157–160
    basis expansions 158
    discretization 157
    Galerkin method 159–160
  set-up 156
fundamental theorem of asset pricing 12
futures price 24

gamma 15, 90
GARCH models
  of implied volatility 150–154
Girsanov's theorem 13, 37, 199
greeks
    → delta, gamma, rho, vanna, vega, volga, theta, 15–19

hedging volatility 40

implied tree
  binomial 67–75
  stochastic 80–83
  trinomial 77–80
implied volatility surface
  → volatility, implied
  estimation of derivatives 57, 98, 103
  least squares kernel smoothing 115–119
  nonparametric smoothing 100–104
  shift factor 141, 173
  slope factor 142, 173
  term structure factor 173
  twist factor 143, 173

in-the-money 20
integrated squared error 105
Itô formula
  multi-dimensional 197
  one-dimensional 196

Jackwerth implied tree 71
jump diffusion 44, 90

Karhunen-Loève expansion 155
kernels 99–100
  Epanechnikov 99
  Gaussian 100
  multivariate 100
  quartic 99

Lévy-process 44, 45
least squares kernel smoothing
    → semiparametric regression
Lipschitz condition 198
local polynomial smoothing
    → nonparametric regression
local volatility
    → volatility, local
local volatility model 67–90
local volatility surface
    → volatility, local
  estimation via local polynomials 57, 87, 98, 103

market
  complete 13, 48
  incomplete 37, 39
market price of risk 13, 48, 65
market price of volatility risk 38, 65
martingale 195
mean integrated squared error 105
mean squared error 104
  asymptotic 104
measure
  K-strike, T-maturity forward
    risk-adjusted 64–66
  risk neutral 12, 38, 49
mixture diffusions 85
moneyness
  forward, futures 20
  log- 20
  stock price 20

Nadaraya-Watson estimator
  → nonparametric regression
nonparametric regression
  bandwidth choice   104–112
  leave-one-out estimator   106
  local polynomial smoothing   57, 87, 102–104, 139
    asymptotic bias   103
    asymptotic variance   103
    derivatives estimation   103
    multi-variate   104
  Nadaraya-Watson estimator   100–102
    asymptotic bias   101
    asymptotic variance   101
    multi-variate   102

option
  American style   10
  barrier   42, 90, 96
  call   10
  European style   10
  forward starting   90
  plain vanilla   10
  put   10
  underlying asset   10
Ornstein-Uhlenbeck process   37
out-of-the-money   20

payoff function
  call   10
  put   10
portfolio
  replicating   11
  self-financing   11
  tame   11
principal component analysis   125
put option   10
  Black Scholes formula   14
put-call parity   14, 15, 191

quadratic variation process   196
quantile-hedging   39

Radon-Nikodým derivative   13, 37, 66, 200
rho   17
risk-minimizing hedging   39
Rubinstein implied tree   71

Schönbucher model   91
Schwarz information criterion   131
semiparametric regression
  least squares kernel smoothing   115–119
    assumptions   116
    asymptotic normality   117, 203–205
    consistency   116, 201–202
    weighting schemes   117–119
semiparametric factor model   160–184
  estimation   164–166
  model selection   167–171
  norming   166–167
  of implied volatility   171–182
  prediction performance   182–184
  set-up   162–164
state price density   18, 29, 52, 60
stochastic differential equation   198
stochastic implied volatility model   91–94
  PDE   94
super-hedging   39

Tanaka-Meyer formula   52, 197, 198
theta   17
time to maturity   20
trading strategy   11, 12

vanna   17, 90
variance
  instantaneous   9
  local   49
variance explained   167
  implied volatility surface   140
vega   17, 56, 117, 118
volatility
  Black Scholes model   9
  constant   9, 14, 19, 34
  deterministic   34–36
  implied   19–26
    as spatial harmonic mean   60, 61
    DAX   33–34
    explanation for   43–45
    forecast   66–67
    interpretation as average   36, 38
    large, small strike behavior   28–29
    link to local   55–62

overview 95
predictor of realized 42–43
slope bounds 27
stochastic 91–94
stylized facts 30–32
instantaneous 9, 49, 60
  Cox-Ross model 84
  deterministic 50, 53, 54, 92
  in local volatility models 49, 50
  in stochastic implied volatility
    models 92
  overview 95
  stochastic 49, 64
  unconditional expectation of 66
local 49–63
  characterization as risk-adjusted
    expectation 51, 64–66
  definition 50
  dual PDE approach 54–55
  Dupire formula 51–55
  implied tree 73, 78
  link to implied 55–62
  mixture diffusions 85–86
  nonparametric approaches 86–88
  overview 95
  parametric approaches 84–86
  slope rule 62, 88
  quadratic 84
  stochastic 36–39, 44
  time-dependent 35
volga 17, 117

Wishart distribution 132